M

DATE DUE	
DEC 0 9 2000	
JAN 0 5 2001	
JAN 19, 2001	
FEB 1 5 2001	
FEB 2 3 2001	
OCT 1 1 2001	
JUN 2 3 2007	

NOV 0 1 2000

THE WAY OF ALL FLESH

THE WAY OF ALL FLESH

THE ROMANCE OF RUINS

Midas Dekkers

Translated from the Dutch by
Sherry Marx-Macdonald

Farrar, Straus and Giroux
New York

Farrar, Straus and Giroux
19 Union Square West, New York 10003

Copyright © 1997 by Midas Dekkers
English translation copyright © 2000 by Sherry Marx-Macdonald
All rights reserved
Distributed in Canada by Douglas & McIntyre Ltd.
Printed in the United States of America
Originally published in 1997 by Uitgeverij Contact,
the Netherlands, as *De vergankelijkeid*
Published in 2000 by The Harvill Press, Great Britain, as *The Way of All Flesh*
First published in the United States by Farrar, Straus and Giroux
First American edition, 2000

Library of Congress Cataloging-in-Publication Data
Dekkers, Midas, 1946–
    [Vergankelijkeid. English]
    The way of all flesh : the romance of ruins / Midas Dekkers ; translated
from the Dutch by Sherry Marx-Macdonald.— 1st ed.
        p.   cm.
    Includes bibliographical references and index.
    ISBN 0-374-28682-5 (alk. paper)
    1. Aging. 2. Life cycles (Biology). I. Title.

QH529 .D4513   2000
612.6'7'01—dc21   00-042677

This translation has been published with the financial support of the Foundation for
the Production and Translation of Dutch Literature

Contents

The purpose of structure is delay.
Eventually, the weight falls,
strain is released.
Years, decades, centuries later
the event occurs.

from "Structures", by Ann Rae Jonas

I

THE STAIRWAY OF LIFE

We certainly know what we want. To move ahead, further and further, especially upwards, until we reach the top. Because it's better up there, even best.

Up there is heaven, down there is hell. A single fall and angels turn into devils. But on earth, too, you have to be on top. Birds fly upwards and defecate downwards. There's no better way of confusing the world than by turning it – or us – upside down. Head and heart belong above, where one thinks and loves, whereas in the lower regions of the body our darker lusts nestle and steaming waste is discharged. While head and heart dream of true love, the groin does nothing but hanker after sex. And when the body, weary of this divided life, finally descends into the soil, surrounded by maggots and centipedes, the soul soars heavenwards like a dove. This riven view of life explains the irresistible tendency of many people to climb mountains, risking everything they are and have; not because the mountain is there, but because it's high. Tops are there to be reached.

People see life as a ladder; a ladder waiting to be climbed, step by step, rung by rung; the higher the better. So you'd expect there to be a lot of pushing and shoving, and elbow-work and sawing through of rungs. In reality, though, at school and in the office, the career-minded get along quite well with their colleagues, and the people down below are no moodier than those halfway up or even at the top. A career needn't be a climb; it's just a *carraria via*, a road for vehicles, and it can lead in any direction.

To reach the top, you have to climb a ladder, but that doesn't mean ladders actually serve that purpose. Social ladders exist precisely to keep you where you are. Thanks to them, everyone knows his place. Hens, for example, have a pecking order. The No. 1 hen can easily be spotted: she's the only one no other hen pecks. But she does peck the others: Nos. 2, 3 and 4, and so on, down the line. A hen halfway down the hierarchy, No. 50 say, gets pecked by Nos. 1 to 49 and herself pecks Nos. 51 to 100. At least, it could work this way. But if a hen were really to peck all the other hens below her, she'd have no time left to lay eggs. A No. 1 hen that pecks all the other hens in the coop would soon be too tired to stay at the top. In reality, hen No. 1 only pecks those directly below her, and hen No. 35 just worries about her immediate neighbours. Thanks to the ladder, you only need worry about the rungs within reach; ladders exist precisely so that you can take small steps, not big ones. Such a system guarantees the stability of the group. Biologically speaking, the rungs of the hierarchy aren't just meant as steps upwards for the ambitious individual, but for the welfare of the whole group. Everyone should be happy, and is, except on the lowest rung. Because that's the good part about this system: almost everyone, however many bosses he has above him, is himself a boss over others below him. Even if you're only No. 95, you still have a higher status than Nos. 96, 97, 98, 99 and 100 – a status worth keeping – which is why Mr Smith meekly accepts his boss's harsh words. A ruler doesn't need to divide in order to rule; the subjugated divide themselves.

Although biological hierarchies defy quick careers, by contrast, careers without biological hierarchies are an impossibility. The fact that the majority of people and animals accept their place in the hierarchy with little difficulty, thereby facilitating the careers of the most ambitious, has deep biological roots. The social behaviour of our simian forefathers still forms the basis for modern management teams, salary scales and chances of promotion. The person who wants to be promoted quickly seeks refuge in the hierarchy of another animal species. If you want to become a boss overnight, get a dog. In a canine hierarchy, you're at the top. That would never be the case with cats. They're not very hierarchical; not because they're too dumb but because they're too unsociable.

Cats don't have careers. People usually do. The only thing a person has to do to get ahead is to age. During childhood you gather the knowledge you need for your first job, and that job turns into a second job. Thanks to the cumulative effects of the inexorable march of time, a person acquires possessions bit by bit: a house, a garden, a husband or wife, a porcelain shepherdess on the mantelpiece, a German shepherd in the kennel. In short, things get better all the time; the top appears to draw ever closer, even though, strangely enough, it never quite becomes visible. Because, by definition, few people experience what it's like at the top.

Attempts to reach the top: that's what makes the world go round, what television is all about, what books and magazines are full of. The women on the covers are young and radiant; in *Sunday Grandstand* young gods and goddesses sprint across the screen, pursuing medals, breaking records, dwarfing past

The Stairway of Life, c. 1640

glories. New houses rise in cities, new cities in countries. You don't see old cars in our part of the world, only in Cuba. Old houses are restored right down to their foundations, old people are banished to the outskirts of cities between the waste incinerators and sewage processing plants. In the visible world everything is new.

That everything is new is in itself new. There used to be more of the old. A century-and-a-half ago most European cities were still enclosed by their medieval ramparts. In the Netherlands, until after the Second World War, you weren't allowed to build with brick near old fortified cities such as Weesp, not far from Amsterdam, because the old line of fire had to be kept open. Not long ago, old and young people lived together in the same house and houses stood next door to ruins. Old and new lived not only above and below each other but also alongside each other. For centuries, people didn't live on ladders but on stairways.

In past centuries, commoners' sitting rooms often had a "Stairway of Life" hanging on their walls. The people depicted on these stairways – drawn, printed, embroidered or carved into cake-boards – often reached the top level not at the end of their lives but only halfway through them. After that, the stairway didn't go up anymore, but down again, like the steps of a cathedral. As eagerly and sure-footedly as people ascended this stairway, they later uneasily stumbled down it, until they finally expired completely. According to Dutch art historian Korine Hazelzet, you were usually given a stairway as a gift to mark a milestone in your life, such as birth or marriage. Until the eighteenth century this custom was reserved for the wealthy, who had the "Stairway of Life" engraved into silver or crystal. Later, the theme became more widespread through popular prints, which, until recently, you could still buy at markets in Italy and Greece.

During the Middle Ages, people didn't yet see life as a stairway or a ladder; they saw it as a wheel, which television has now made familiar to us as the Wheel of Fortune. But in the sixth century, instead of today's "has-been" quizmaster turning the Wheel, it was the blind goddess Fortuna. She did so not to satisfy man's avarice but to expose it. People knew this story from the medieval equivalent of television, the stained-glass church windows, to which their attention strayed Mass after Mass. These depicted many a king toppling from his throne and were instructive and entertaining at the same time.

If you imagine life as a circle, then the infant is followed by the child, the child by the adult and the adult by the old man; but after the old man comes the infant again. *Les extrêmes se touchent.* Old people and children have always had a special bond. Grandchildren usually get along with their grandparents as well as their grandparents get along with them. What one cannot yet do well, the other can no longer do well. While one crawls because its legs are still rubbery, the other walks bandy-legged because its legs have become brittle. Young and old are literally and figuratively more earth-bound, more animal-like; sometimes both wear nappies; they're friends in fate. They both face the unknown. To the pious believer, a cyclical life was heresy. The essence of a Christian life was that it had an earthly end, a fulfilment, at which time everlasting life began elsewhere. The "Stairway of Life" served literally as the writing on the wall, to warn you during every phase of life, of the temptations Satan had in store for you. During the sixteenth century, the stairs were filled with rakes, drunkards and misers. No one ever got into heaven this way of course; that was obvious from the start.

What made the "Stairway of Life" so appealing was its perfect symmetry. Every step you went up, you also had to come down again; the descent was given as much attention as the ascent. Truth would often be bent a little to achieve this. In the interests of symmetry and round numbers, life was made to end at 100, so the peak came at 50. For most people in those days, 100 was less attainable than it is for us. But it wasn't impossible. The shockingly low average age was due mainly to the high infant mortality rate. Whereas today you probably die because you're old, in those days you died because you were young. The most common length for coffins was less than a metre. But if you survived the measles and the poxes, you could even reach a ripe old age, like Michelangelo, who lived to 89. So how was 50 sold as the peak of life? This can be seen in the symbols that were used to decorate the different steps in the "Stairway of Life". You were given a lion on your

The Wheel of Life.
Wood engraving, c. 1480

fortieth birthday, while a fox was reserved for your fiftieth. Perhaps the old fox's greater cunning was supposed to compensate for the first loss of physical stamina? The peak of a man's life was at the moment when ascent and descent kept each other in balance:

> 10 years a child
> 20 years a youth
> 30 years a man
> 40 years mature
> 50 years the turning point
> 60 years the downward slope
> 70 years grey-haired
> 80 years ne'er wise
> 90 years children's scorn
> 100 years God's grace

We take it for granted that everything is divisible by ten, but in medieval times, people didn't. They divided everything by seven, in accordance with the seven deadly sins, the seven good deeds, the seven liberal arts, the seven days of the week and the seven loaves of bread from Christ's miraculous feeding of the 5,000. Consequently, fifteenth-century wheels of life comprised seven phases of life, often with the same attributes that would later reappear on the ten-step stairway: a cradle for the baby, a hobbyhorse for the child, a falcon for the adolescent, a uniform for the young man, money for the middle-aged man, a cane for the old man and a grave for the dying man. The seven phases of life were also associated with heavenly bodies, seven of which had already been identified in antiquity. Ptolemy had taught that children were influenced by the moon, which, after all, also changed its appearance quickly. Mercury planted the seeds of knowledge in you until you were 14, after which Venus made you lose your head again. Via the Sun, Mars and Jupiter, you eventually ended up with Saturn, who made you die a miser.

The logic of this system escapes us, but medieval people weren't looking for logical explanations; they were seeking harmony. The more harmony they found, the better they could understand the world. That the world was harmonious was unquestionable: the Creator wasn't a charlatan. So connections were never coincidental. If you could divide two different

The four body fluids in classical medicine: blood, phlegm, yellow bile and black bile things by seven, they must be related. But twelve parts – the zodiac, the months, the apostles – or four would do just as well.

Quartering comes so instinctively that we still do it today. The compass is divided into four directions, the day into four parts, the year into four seasons. Life is too. After the spring of life comes the blossoming, followed by the autumn and the winter. The analogy was taken from plants, most of which pass through the four phases of human life in one year – from seed to plant to blossom to death. The ancient Greeks delved even deeper. Pythagoras explained all material in terms of four elements – water, fire, air, earth – that existed in four states – hot, cold, wet, dry. According to classical medicine, these elements could be found in the fluids of the human body: blood, phlegm, yellow bile and black bile. Health was nothing more than a proper balance between these fluids. If the balance was disturbed, the doctor would restore it by blood-letting. Naturally, this system was also age-bound. Children, because of their vitality, were cheerful and recovered quickly from disappointments; later, because of an excess of yellow bile, they became ill-tempered and pouted a lot. That adults were so slow, unexcitable and rancorous was ascribed to black bile. And excessive phlegm made the elderly so melancholy that they never forgave anyone anything anymore. Incidentally, this grim picture of old age contradicts the oldest summing up of life's phases in our culture, dating from 2,600 years ago and attributed to the Athenian lawgiver Solon. According to him, in the sixth, seventh and eighth of life's ten phases, people were so wise they no longer wasted their time on useless things and were able to formulate their profoundest insights most succinctly. Apparently absolutely everything was better in those days – even old age.

The "Stairway of Life" has disappeared from our sitting rooms and as a result we've lost half of our lives. Old age is no longer good or bad, it's simply denied. There's no place for old people anymore. They dress young, imitate the way young people talk, have their hips replaced, take cruises. Of the ten steps, only five are left, and these are made to last as long as possible, until man suddenly falls off the highest one. The twilight years have even largely

A modern
"Stairway of Life"

disappeared from biographies. Endless pages are devoted to the early years of famous men and women, like Freudian preludes to the great and moving lives that will follow. But everything that happened after the great deeds is dealt with summarily, in terms like Mother Goose's "lived happily ever after". Where did they go, the world leaders and sporting heroes of yesteryear? Of course we know that Churchill took up painting, Wałesa applied for a job at his old shipyard and countless cabaret singers pined away in draughty garrets. But we generally prefer famous lives to end at their peak; no easing off. In that sense, John Kennedy, John Lennon and Princess Diana had a strong sense of duty. And dictators aren't even supposed to come near pensionable age. Adolf Hitler, Ceauşescu and Napoleon, together with many kings and emperors, never got to go down life's stairway.

Their empires suffered the same fate. Nothing is so short-lived as a 1,000-year empire. Whereas in the past, empires tended to disintegrate slowly, languishing in centuries-long decline, today new empires simply take the place of the old ones. All trace of Adolf Hitler's 1,000-year empire was swept aside in one decade, supplanted by the Christian Democratic welfare state. In those days, arch-rival England still ruled over an empire of more than 800 million people. George VI was not only the Emperor of India but also the King of Australia, Canada, South Africa, New Zealand and Newfoundland. Today, the Queen possesses only a few remote rocky promontories and clusters of poverty-stricken islands inhabited by fewer than 170,000 tea-drinking locals. "Never," wrote the British writer Simon Winchester when Hong Kong was handed over, "no, never did a world empire fall apart so quickly and so quietly as that of the British colonies." Communist regimes no sooner fell than statues were dragged from their plinths, hammers and sickles were torn from newly unemployed hands and the whole body of thought was made suspect. New ideologies beckoned. Better ideologies of course, because new is better where ideas are concerned. As lifeless as "NEW!" has become in advertisements, it flourishes like never before in our view of the world. But it wasn't always like that.

During the sixteenth century, the "Stairway of Life" was populated with reprehensible specimens, while during the seventeenth century everyone on them had become respectable and good. Even the dog, who initially stood for the grouchiness of the old miser, had been re-educated. He now guarded the money that the virtuous old lady had frugally scraped together. Good and bad look very different today. For us, the ascending part is good and the descending part – inasmuch as we dare look it in the face – bad. We prefer to stay young throughout our whole life, and then, when death really does become inevitable, to exchange the fleeting for the everlasting, with as little old age as possible. Death itself isn't such a problem. It has become such a hot item over the past few years that it echoes in our ears. The focus has usually been on how to get life to turn into death as quickly and painlessly as possible – that is, without any real old age. To achieve this, people even resort to the impossible: suicide carried out by someone else – euthanasia. Insured to the hilt, after heartfelt farewells to our loved ones and with a euthanasia pill under our tongue, we hope to slip away unnoticed into the pink satin-lined coffin. We tolerate the fact that a highly unpleasant adolescence was introduced between childhood and adulthood, but we don't accept the senile stage between adulthood and death.

Animals have less difficulty with old age. Cats, for example, like us, have only one life, but they divide it up better. By nature their stairway of life is a wide plateau with one little step up on the left side and one little old-age step down on the right. It's easy to see when kittens are young – how cute! – and tomcats are old – how touching! – but during the many intervening years it's very difficult for an amateur to guess a cat's age: is it two? six? ten? Cats enjoy the prime of life for more than four-fifths of their waking hours. Compared with them, we hardly even get started on a productive life. If you finish school at 20 and retire at 60, less than half your life remains. And you sleep away one-third of that. This in contrast to cats, who sleep away no less than two-thirds of their lives.

Because of all their naps, cats don't make the "adult impression" you would expect of animals that are adult for most of their lives. Moreover, until a ripe old age, cats behave like kittens when we're around. They're always eager to mistake a dangling piece of string or a spool of thread for a mouse. The playfulness kittens need in order to learn how to hunt stays with them throughout their lives. The same cannot be said for human children: they grow up, and then they're no use to us anymore. If you really want children, before you know it, you're saddled with more adults. Cats stay children all their lives. And it's precisely because their stairway of life is so different from ours that they're so well-suited to living with us.

In the wild, cats are old for an even shorter time. Street cats usually reach the age of two. Wild cats, like tigers, face death the moment they can no longer keep up with their prey or when they miscalculate their leap. Some old tigers then turn to the easiest prey of all: man. But to find really old tigers, you have to go to the zoo.

They say you only have to multiply a cat's or a dog's age by seven to estimate

The stairway of life is often
drastically shortened in nature

its age in human terms. Well, that's not true. Because of the totally different nature of their stairway of life, cats or dogs become at least 14 human years old during their first year of life. In the case of other animals, the stairway of life is even more distorted. Birds, for example, appear not to age at all. As long as they're not dying, it's almost impossible to tell how old they are. That time also takes its toll on them is reflected mostly in their reproductive skills. As they get older, they lay fewer eggs and their eggs are less likely to develop into chicks. In the same way as the great female titmouse lays fewer eggs after about five years, the male at that age becomes less successful in defending his territory. The result is that the pair occasionally miss a year of nesting.

The childhood of a great titmouse is as rich – the warm nest, the not-to-be-scorned delicacies that Mum and Dad spoil their little darlings with – as the early years of ducks and chickens are lean. In their world, ruthless child labour is the norm. No sooner have the young emerged from their shells than they have to walk, swim and fend for themselves. But *Adactylidium* mites beat all known records as far as the absence of a childhood is concerned. They're born orphans. The young mites leave their eggs while still in their mother's body and then eat their way out. As a consequence they have little time to be young; their mothers, on the other hand, have no worries at all about becoming old.

The stairway of life is often drastically shortened in nature. While today, human beings tend to die old, animals usually die young. As once was true of people, the average life span of any animal species, especially shortly after birth, is cut short by death. The younger animals are, the earlier they die. Danger awaits them on all fronts. Shortly after birth, tadpoles and guppies run the risk of being eaten by their own mother. Even when I see them on television, the sight of newborn green turtles (*Chelonia mydas*) – better known for their role in soup – making their mass exodus to the sea deeply moves me. Hardly out of the egg and still recovering from their struggle with the shell, the baby turtles scuttle as fast as their little legs can carry them across a beach full of baby-green-turtle-eaters. For the snakes, monitors and predatory gulls, this D-Day in reverse is an annually recurring feast. How eagerly they sink their teeth and claws into the still-soft shells! The babies disappear, thrashing, into beaks and jaws; the odd one continues its journey on three legs; more than half never reach the water, to say nothing of the soup. Even the winners of this unparalleled competition aren't safe; at sea, barely one-in-ten survives even the first week. It's like a scene from the Last Judgment. You'd have to be a very bad cameraman not to make an epic story out of it.

Yet these baby turtles are well-off compared with acorns and chestnuts. Of course, it's a miracle that a whole tree can grow out of one such fruit, but for every oak tree in the forest, 10,000 acorns have been produced in vain. They lie rotting on the woodland floor, a discarded investment. If an average factory were that wasteful – throwing away 9,999 cars out of every 10,000 – it would be bankrupt before the week was out. Once finished, though, a tree lasts much longer than a car. Even compared to humans, oak trees have the nearest thing there is to everlasting life. During their long existence they're capable of producing more acorns than Mother Nature could ever manage to waste. But they don't have everlasting life. Eventually they produce fewer acorns of a diminishing quality. Fruit farmers are familiar with this phenomenon from their apple and pear trees. After a few years the harvests shrink rapidly. As for the newer varieties, the young trees are replaced by even younger ones long before they're fully grown themselves. Annuals don't even get that far. Much to the delight of garden centres, they're only around for one season. They germinate, grow, produce seeds once – and die. If the seeds don't find the right home, another plant has to be bought the following year. To grow a whole plant, with everything that entails right up to the finest anthers, and then to throw it away after using it once: there's no end to nature's wastefulness. People go one step further. They grow plants that can do nothing but germinate and grow. These plants are cut off before they can produce seeds – to be put into vases. The only thing the flowers can do there is die. Some are really good at it too. That's what cut flowers are paid to do; it's their calling. People delight in watching them die. They make them into large bouquets or give them to other people so that they too can share in the pleasure of the flowers' terminal phase. Many bouquets accompany people on visits to hospitals. All patients feel better after watching their flowers die before they do.

Annuals are lazy. They dislike the prospect of having to tide themselves over the winter or the dry season. The more baggage you accumulate, the more you have to discard. Trees shed their leaves before winter begins, weeds go back to their roots. Annuals are even more conscientious. They die altogether, putting their survival in the hands of their seeds. These are small enough to slip through winter's net and are much more prolific than plants could ever be. Many insects see something in this method too. The journey to warmer climes is too demanding for them, and hibernation in the compost heap too dangerous. But why should they bother sitting out the winter? Their bodies were built to last for only a few weeks or months anyway. They lay their eggs, which are well equipped to withstand heat and dryness, before winter begins. After laying them, there's no reason for creatures as delicate as moths to continue living. With their mouths having to degenerate, silk-moths can't even eat anymore. But what does it matter? That's not what moths are for. Caterpillars take care of that.

Organisms that only reproduce once in a lifetime are called semelparous. This is in contrast to human beings, who are as iteroparous as rabbits. You don't need a short life to be semelparous. However old a salmon or an eel may become, it makes only one journey to its spawning grounds, where, after depositing its eggs, it dies from exhaustion – not just because of the long journey through the ocean and up the river, however tiring that may have been. Salmon are already no longer themselves by the time they reach the mouth of the river. The males develop thick hooked jaws and hunchbacks. After spawning, they become covered with a fungus that begins to eat away at them, while inside, one organ after another starts to give way. Hormones no doubt play a role here. The same hormone system that drives the creatures to spawn also causes their death. Sex and death are faithful allies. Seventeen-year-cicadas, otherwise known as periodical cicadas (*Magicicada septendecim*), live chastely underground for 17 years until they emerge from their cocoons to claim in one night all the sex they were meant to have in a whole lifetime. It's a long run for a short jump, but a productive one all the same. After all, 17 is a prime number. It has no factor other than itself. Predators that have their sights set on these cicadas should themselves have at least 17-year cycles if they're to benefit from this mass nuptial celebration. Although eaters of 16-year-cicadas would live in great uncertainty every 2, 4, 8 or 16 years, they would also have the chance of an unforgettable feast. Which is precisely why 16-year-cicadas don't exist. Whoever becomes available as prey at predictable moments has to take part in the struggle more frequently and in larger numbers. You can set your clocks (or thermometers) by mayflies. When, as usual, they rise up in their massive nuptial flights – in the right temperature on the right day at the right time – and then later cascade down again exhausted, ponds and lakes teem with fishes having the time of their lives. Some fishes have even adjusted their mating season to accommodate this. Because of this phenomenon, mayflies don't even live out their one day. A perfect symbol of ephemerality, you might say. Yet nothing could be further from the truth. Mayflies live not for one but for 100 or 1,000 days. Of these, though, 99 or 999 are spent in water. As with 17-year-cicadas, their stairway of life is one high, narrow step up, with one low, very wide step down. Pity isn't appropriate here. Creatures with such long larval phases have, after all, the key to the most coveted of all secrets. Mayflies have eternal youth – with the exception of that one hellish day.

It's good to have a larval stage. Young creatures have different tasks in life to those of their parents and so they function differently. In the first place, they need to grow, while old creatures need to make young creatures. Caterpillars are nothing more than eating machines. They consist mostly of jaws and intestines, which they use to ravish whole forests. They don't need wings to do this, and penises and vaginas would just get in their way. Butterflies, on the other hand, are winged genitals. They spread their species elegantly and nimbly. To fuel their flying and reproductive systems, they cash in on supplies from their previous existence as caterpillars, or coquettishly lick up nectar here and there. In the case of barnacles, the roles are reversed. Because, as adults, their calcareous shells are attached to rocks, they use their larvae to propagate. The larvae allow themselves to be swept along with the currents until they find an appropriate place to settle far away from home. For reproduction they have penises at their disposal many times the length of their own bodies. Despite their immobility they thus always manage to penetrate a neighbouring female. Just to be sure, though, barnacles are both male and female at the same time.

The most common animal species with a larval phase is man. You wouldn't make a young mother very happy by calling her newborn baby a larva, even though it has all the features of one. Babies can do hardly anything: they have to be spoon-fed to eat; they can't walk, stand or even sit. True, their

The only interesting thing about a baby is its future

heads wobble a little, but aside from that, what they resemble most are patients with spinal cord lesions from the neck down. All the muscles below the neck are extremely unpredictable, especially the anal one. Their crooked little legs suggest rickets and their speech is severely lacking. Rehabilitation is out of the question because the poor little things have never been habilitated.

They have to be pushed around in prams. What greater failures could you ask for? Any objective mother would tuck her messy little piece of work out of sight, cancel maternity visits and sue the midwife. But mothers aren't objective. They happily announce their new arrival, and visitors don't let on. To divert attention, they ask whether it's a boy or a girl. A stupid question. What's lying in front of them is neither. As for all the things that distinguish boys from girls – playing cowboy, hitting sisters, seeing who can pee the furthest – the little mite doesn't have a clue about any of them. Aside from the odd minor detail about the way nappies are filled, there's no difference at all between males and females at this stage. What visitors really want to know is what the little mite will become later, not what it is now. The only interesting thing about a baby is its future.

Looking at newborn babies for the first time can make you have serious doubts about the survival of the fittest. If those toothless little maggots, those drooling cocoons, those pale-faced grubs are the fittest, why do all other animals have access to teeth and claws, or venom, mimicry and quills? And, if they're the fittest, why in God's name do they survive? It's for one reason only: because, naked and pale as they are, they can do everything babies are supposed to do. Babies can no more supply their own milk than cats their own biscuits. But what does it matter, as long as we're stupid enough to feed them? Babies can get their mothers' attention more efficiently than fathers can get a waiter's in a restaurant. If, by chance, mothers don't react to the heartrending little glances or clutching little hands, babies just start screaming at the top of their lungs. This is when we pay the price for the fact that human beings originally inhabited wide, open plains. The screaming is too loud for a three-room flat, but it's effective all the same. Mum and Dad can't get there quickly enough to fill their little mites to the brim. Eating is what it's all about, for human as well as animal larvae. And so they shoot up: little hands become big hands, little legs become big legs, and muscles begin to understand what they're meant to do. The larvae truly begin to resemble human beings. Do babies benefit from all this? Of course not. You may become a better person from growing, but you become a worse baby. As toddlers, children have already forgotten much of what they could do as babies: be born, live from milk alone, breathe and drink at the same time – they can't do any of it anymore. The easy life ends all too soon. Little children aren't nearly as efficient at manipulating their mothers as they were as babies. Instead, they peer jealously into the cradle of the newest arrival. What can their parents possibly see in that pale-faced dumpling?

The transition from baby to toddler and toddler to schoolchild is gradual, but for a schoolchild to become an adult, it, like all larvae, must undergo a radical transformation: adolescence. Covered with pimples and falling over its own legs, an inquisitive little monkey changes into a sex-crazed human being.

Children's books are cast aside in favour of cheap romances, pulp fiction and porn magazines. From the moment an adolescent first falls in love, the peanut-butter champion of the world can no longer stand the sight of his favourite foodstuff. Gone are the days of sitting quietly in a corner. The larva emerges from its cocoon, makes a noisy début, frequents discos, ventures out into the world. As a new adult it goes forth and multiplies. That's what adults are for. But first the child has to be done away with. Childishness won't get you very far during a romantic dinner for two. Where is the child you once were? Where is the gangly adolescent, the shy schoolboy, the little girl with braids? They are – like so many people in your photo album – dead. Dead but not buried. Eaten up. In the same way as a butterfly can only be born after it has first eaten the caterpillar it once was, so you have destroyed the tissues of your child's body, cell by cell, and rebuilt them as adult tissue. A child's greatest enemy is the adult that will grow out of it. Adolescence is nothing more than the war between these two. In adolescence you succumb to your own gunfire. In this way you rehearse for your second death 50 years later. Death is nothing other than delayed execution for infanticide.

Few children with the rare ageing disease
progeria live beyond 15 years old

Every species has its own stairway of life. You have to climb it: step by step, missing none, never going back, at best being hurled off when you're only halfway up. What varies is the tempo. One species skips up the steps like a child on the steps of an Amsterdam canal house, the other crawls through life like a tortoise. Illness speeds things up. An extreme example is progeria. Children with this rare disease change into old men before our very eyes. They quickly wither away. Wrinkled, toothless, with prostate problems and a ramshackle heart, they die as teenagers. There's no remedy. If there were, we'd have a weapon against all ageing. For the time being though, all progeria patients have to live with the certainty that they'll no more reach 20 than we will 100 – and with the uncertainty of whether that's really five times as bad. While a handful of researchers have devoted themselves to studying a handful of progeria patients, most researchers are interested in how they can slow life's journey. Compared with other animals, we're already very slow in our development. This has to do with our large brain. It needs time to grow and that can only happen if we stay young for a long time. If you compare an adult human, not with an adult but with a newborn chimpanzee, you wonder why the theory of evolution wasn't discovered much earlier. Much of what distinguishes the other apes and monkeys from us – thick eyebrows, protruding jaws, hirsuteness – is still absent in baby chimpanzees. Human beings are apes that retain many of their youthful features until an advanced age. This is why we stay so hairless all our lives and why our brain continues to grow for so long. We come into this world as tiny embryos and leave it as big babies. "The missing link between apes and men", Konrad Lorenz said, "is us." Thanks to our long childhood, we become sexually ripe at a late age. As we then have to take care of our own slow-growing larvae for many years, we must grow old in order to preserve the species.

The classical "Stairway of Life", with its uniformly high and beautifully symmetrical steps, seemed to suggest that we grow old gradually, at a steady walking pace. But nothing could be further from the truth. By the time you get your first grey hair and are convinced you're on the slippery slope, the worst is already behind you. You did most of your ageing when you were very young. With their bald heads and toothless gums, newborn babies not only resemble old men, in some senses they *are* old men. They age more rapidly then than they ever will again. Not only is their body feverishly constructing, it's also feverishly deconstructing. Babies are like Rotterdam, where they pull down buildings as soon as they've finished putting them up, only to erect new ones in their place that are destined to lead equally brief lives. In adult bodies, construction and deconstruction take place much more slowly, as in the older quarters of Amsterdam. But tissues are not made of stone. Their regenerative powers are forever decreasing. In old people the vigour's gone and they die. But it's not their fault. It was their younger body, not their older one, that used up all the vitality. During a single month in a baby's life, the ability to replace old cells with new diminishes more rapidly than during one whole year of an old person's life. "Our notion that man passes through a period of development and a period of decline is misleading," wrote Charles Minot as early as 1908. "In reality we begin life with a period of extremely rapid decline, and then end life with a decline which is very slow and very slight." An old person shuffles towards death inch by inch, whereas a baby gallops towards the grave. An older person is slower at everything – even dying.

By the time he reaches the end of his stairway of life, a man hears footsteps behind him. For the umpteenth time a new generation is climbing the stairway, full of expectations. Little do they know. But one day the footsteps will cease, one day they will be finished, one day the old man will no longer be replaced by the young man. This will be when mankind has died out. It will be a great relief to animalkind and plantkind, I think, but we humans don't much like the idea. "Dying out is a very radical way of dying," laments one Dutch writer. But I'm on that very path myself. I have no children. For the first time since Adam and Eve, someone in my evolutionary line has failed in his duty. I'm old enough to have had children, but I didn't. I'm not worthy of the chosen people. By the chosen people, I mean everyone who's alive. We're exceptionally privileged. Imagine if you'd been a miscarriage. Of course, then you wouldn't have existed. But the same would apply if your mother had been a miscarriage. And if your great-great-great grandmother had died before giving birth to your great-great grandmother, all those mothers between you and her would never have existed. Time and again, your foremothers managed to be in that one egg cell that was fertilized by one of those many millions and millions and millions of sperm cells. The chance that you exist is nil. And yet here you are. And here am I! The result of a series of miracles. And yet I am too uncaring to prolong the series with yet another little wonder. I'm severing a line that is hundreds of thousands of years old. If that same behaviour ever applies to everyone, mankind will be finished. Lived out. Died out.

Is this a bad thing? After years of faithful service, why shouldn't an animal species be allowed to disappear from the face of the earth? No one knows. All the arguments against the extinction of animal species can be easily refuted. Not even life's beauty or our respect for it adds enough weight to the scale to justify the existence of the World Wildlife Fund, the Friends of the Earth, The National Trust or Greenpeace. The beauty argument – "You wouldn't want Rembrandt's Nightwatch to be destroyed, would you?" – is inappropriate, because there are so many more animal species than Nightwatches and in so

many more forms. Even if 100,000 species were to be killed off, there would always be more left to enjoy than we could ever find time for in our whole life. And the motto "reverence for life" isn't quite so viable if you take one random example: the tapeworm perhaps, or the AIDS virus. So why do we get so angry when yet another species is done away with? Why are we saving the whale and making more generous donations to the plaster panda than we ever did to the golden calf? Not because we mourn all those lost animals – do you miss the

Do you miss the dodo?

dodo? – but to spite our fellow human beings, who have those losses on their conscience. It isn't the broken toy or the lost postage stamp that makes us angry, but the naughty child who broke or lost it. In itself, extinction only benefits the animals. They become much better known because of it. Who would ever have heard of the quagga, the very recently extinct relative of the zebra, or of the Tasmanian wolf, if they hadn't become extinct? Extinction enhances your status. Pandas are heading in the right direction. The only thing that bothers us is the avoidability of it all, the chance that later we'll have to say "If only we'd . . . ". And that's precisely the weakness that nature conservationists exploit. That's also why nature conservationists won't die out quickly. Everything else we value will probably walk the gangplank first. Extinction is inevitable. Transitoriness belongs to the same family as "the losing battle".

The extinction of the dinosaurs has had the deepest impact. That a group of animals so large – in all senses of the word – could disappear from the face of the earth after 100 million years! The mourning of their loss is amply compensated for, though, in romantic speculations about Lost Valleys and Jurassic Parks where dinosaurs are still believed to roam. Wouldn't it be wonderful if there were still some around, if only from a few species, so that we could observe them alive and well in action? We have that privilege, albeit with elephants instead of dinosaurs. What we call elephants are the meagre remains of what was a much larger animal group, hundreds of species of which once populated the earth. When the last elephant eventually dies, not only will a species have disappeared, but a whole concept, a whole building plan along with it. The last elephant on the last step of his stairway of life represents the last of the order of proboscideans: a full stop at the end of 50 million years of evolution. In the animal world, it's usually the specialists that reach the last step. They've focused so intensely on one strategy that there only has to be the slightest change in circumstances for them to breathe their last. In the case of elephants, that strategy is being big. It has helped them against all enemies except man. But it can't help them for much longer. The end is in sight. Hurry! Come and look before it's too late!

It's equally possible that we might only have known elephants from fossil records. Then we would have speculated on how they might have looked if they'd been allowed to evolve for longer. At least that's certainly what we're doing today with dinosaurs. What would they have looked like if they'd been given another 75 million years? Visions appear in our mind's eye of modernized dragons, science fiction monsters, perhaps even giant family pets suited only for mammoth transport. Reality outstrips even this fantasy. Modern dinosaurs fly. Millions of them crisscross international airspace every year. Birds are nothing more than small flying dinosaurs. If you don't believe it, take a look at the scaly talons and reptilian necks of vultures. Could someone have predicted *that* 100 million years ago? No. Dinosaurs might easily have taken to the water or opted for a life underground. They might have turned gold, or bright purple with polka dots. Evolution is unpredictable. The theory of evolution is all hindsight: history. Things are the way they are by sheer chance.

The future of our species is as uncertain as that of the elephant. We don't know which step of the stairway we're on. On the one hand, we're a comparatively young species; a few hundred thousand years is nothing in a world where most species live for a few million years. On the other hand, we're as specialized as the elephant; all our bets have been placed on one organ: our brain. Thus far it has served us well, but it can't cope with unexpected new epidemics. Like all super-specialists, we have one foot in the grave. Will we survive that long? Only time will tell. But as difficult as it is to predict the future of the human species, it is easy to predict the future of a human generation. Discounting exceptional circumstances, we can calculate on the back of a cigarette packet what, say, the age distribution of the population of the United Kingdom will be in 10 to 25 years' time. That's why it's so irritating that the growing ageing population is talked about as though it were a problem. Anyone in the 1950s could have predicted the grey boom. A baby boom is always followed by a grey boom; it's that simple. We've had 50 years to take steps to deal with it. The fact that nothing was done isn't due to short-sightedness but to our having looked the other way. We don't want to know about the last phase of a whole generation. At school our children are taught about birth and growth, but never about death and decay. Just as you never see pictures of old dogs in books on pedigree canines, the illustrations in their biology books show people in the prime of life. In a greying world, our children are raised on magazines and television programmes full

The *Übermensch*

of blondes and brunettes. We want nothing to do with moving backwards; we only want to move forwards. We prefer winners, not losers.

It can't just be a coincidence that depictions of the "Stairway of Life" disappeared from domestic walls during the nineteenth century – the century of progress. The whole world was seething with expectations. Everything everywhere was always getting better: horses became locomotives, heathens Christians, towers higher, the world bigger. While trees didn't grow sky-high, factory chimneys did. Animal protectionism and Esperantism were discovered, with the theory of evolution as the jewel in the crown. Despite Charles Darwin's actual views, evolution was generally seen as progress. Weeds and worms developed into fishes and birds, and eventually, apes and monkeys became human beings and Europeans. Who knew what was in store for us? Maybe even Europeans could be improved upon – with the help of eugenics. If you could make pedigree cattle from aurochs, the wild ancestors of our domestic cow, and well-behaved dogs from carnivorous wolves, there must be room for improvement in human beings as well. Only much later – after Adolf Hitler tried to hasten the birth of the new man by eradicating a number of old races – did the *Übermensch* turn out to be more carnivorous than the "unimproved" version. A short time later, it became clear that the world could only take so much progress – because of the environment and such things.

There were vague mutterings about regression. From being the pinnacle of creation, human beings had become the only creatures to foul their own nests and do away with members of their own species on an unparalleled scale. Lack of natural selection had weakened and corrupted us. Old people once again jumped at the opportunity of emphasizing that everything was better in the past. Was the human race finally in its demise? Had the West sunk to the level of Sodom and Gomorrah? Would we be ruined by decadence? Were we digging our own graves?

Of course not. We're probably just as good and just as bad as we were in the days of the Romans and the Cro-Magnons. Most people are more affected by two pints of beer than by 50,000 years of evolution. And that's precisely the point about evolution: it's not going anywhere, it doesn't want anything and it doesn't give a toss about progress. Human beings are no better than whales, and whales are no better than the lice on our head. Millions of years of evolution haven't made the world one iota better or worse. It's a difficult idea to digest but, in biological terms at least, life has no goal. Our culture is glutted with goals, strategies and challenges. We want nothing better than to move forwards. But there is no movement forwards. There's not even movement backwards. There's only movement.

And that's what we have to make do with.

2

ROMANTIC RUINS

They say there's a place somewhere on this earth where old and life-weary elephants go to die in peace. One last time their trumpeting rumbles through the depths of the jungle, one last time a tremor takes possession of their bodies, then the great beasts fall to the ground with a dull thud. No one hears them, but the vultures see them and the hyenas smell them. Layer by layer the cadavers are dismantled until they rise up, like white-toothed combs, out of the fields of the ruins of the bones and carcasses of their predecessors.

Unfortunately, such an elephant graveyard doesn't exist. But this didn't prevent me from being deeply moved during a recent trip to South America, when I stumbled unexpectedly upon something that can only be described as a locomotive graveyard. The cadavers of a dozen or so steam locomotives – the state of decay was so advanced I couldn't tell exactly how many of them there were – lay there rusting away. Tropical lianas were fighting to strangle axles, saplings shot up out of funnels, one tender seemed to be transporting flowers instead of coal. Everything was richly decorated with stray bolts and wheels. Streaks of oil evoked traces of blood. The sheer size of it heightened the drama. There's always something especially tragic about the death of a giant. It's true for dead elephants, beached whales, King Kong, Ralph Richardson, even for man-made things. So much dies so suddenly.

The experience was both oppressive and liberating at once. What made it so beautiful wasn't only that the locomotives had been spared the fate of the blast furnaces – being melted down into garden tools for the benefit of some lady of leisure in Surrey – but that they'd been left to stand and rot in their

The locomotive graveyard

own good time. Don't get me wrong. Of course I think buses and trains should be restored, but I think it's more important that somewhere on our overly organized planet – and I have no intention of telling you where – there should be a place where they can truly rest in peace. It has something to do with dignity. In locomotive graveyards, locomotives don't have to die; they're allowed to pass away. God rest their souls. For now and for ever more.

Just try dying in the Netherlands. Nothing, ever, anywhere, can die a natural death there anymore. Still warm from their final run, old locomotives are put out to pasture according to a schedule. Nuts and bolts are collected as if they were evidence for a murder trial and then polished to become *pièces de résistance* in those mausoleums known as railway museums. There the locomotives stand, as unauthentic as can be, too new to be old, yet too old to be new – sterilized, social misfits. Somewhere, beneath all those layers of varnish, is supposed to be the real locomotive, but you certainly can't see it. How can such an anomaly ever evoke anything in anyone? As readily as I can imagine the engine driver standing in such a Bolivian wreck or hear the fire roaring on the grate or smell the stokers' sweat, it is difficult for me in railway museums to envisage anything but the men restoring it. The links with the past have been polished out of existence.

The aim of restoring something to its former glory is as futile as it is human. A lot of glory is simply at its best if it's decaying. Decaying is living. A wreck is less than the locomotive it once was but also more. Different shapes appear, different feelings are evoked than the ones we knew before. The best proof can be found in the ruins of ancient Greek and Roman temples. Somehow, three fallen columns, with the tympanums that once adorned them lying at their feet, radiate more dignity than an entire basilica celebrating high Mass. The sentiment that the ruins of a castle can stir up has nothing to do with the pride a knight must have felt when his castle had just been completed. While the stones of a new castle or palace are all in exactly the same state – neatly finished, perfectly aligned – in ruins, each stone is different, marked by life in its own way, just as every old tree trunk has its

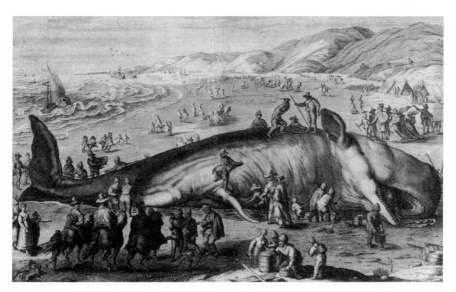

There's always something especially tragic about the death of a giant

own character while saplings are a dime a dozen. Archaeologists who try to imagine what it must have been like to wander about in the halls they excavate, or to look at the paintings as the kings and emperors once did, should be glad their dreams do not come true. Whoever loves whole buildings should be in Milton Keynes, not Herculaneum or Pompeii.

There's more life in ruins than in castles. In the same way as mice come to life a second time a few days after they have died, thanks to swarms of maggots, decay is change and change is life. Stones crack, columns slip off their pedestals, roots pry open walls, steam boilers develop holes, little boys steal manometers. In railway museums, or The Tower – "Opening hours: Mon. to Fri. 12–5, Sat. and Sun. 10–5. No dogs allowed" – all change is stopped in its tracks. Nothing is more piecemeal than a museum piece. The only places left where you can still see life after death are slums, junkyards and scrapheaps. There, you see one of the main features of ruins, once described by the sage of Dutch journalism, Henk Hofland, as " . . . the entrails of the original slowly emerging. In a building, for example, you see the stairways, the water pipes, the sewage system; in a car, the wiring, the motor; in a plane, the frame and ribs of the fuselage and wings; and in a large ship, all of these things."

"A large ship": the spectre of the Titanic instantly looms up. Even though she was already world-famous by the time she was launched – a triumph of technique, the pinnacle of luxury and unsinkability – she only really became immortal after she sank. Had the Titanic ever lived up to her claim of sea-worthiness and reached her intended final destination – the salvage yard – a film of the same name would never have been made 50 years later for many times the costs of building the original ship. If the horrors of this shipping disaster leave you cold, you must have almost no imagination. I wouldn't want to know a person who has never tried to imagine what it must have felt like – first the disbelief, then the panic, the pushing and people's true nature coming to the surface. The image of this giant ship in her death throes – in the end, only the rump protruding vertically out of the sea, the musicians tumbling out of their chairs – competes with a vision of what was happening under water: the prow in the sand, the swollen passengers' faces at the portholes, a fish biting a toe, the coal still stored pointlessly in the bins. A ship meets her fate. With every extra year that the *Titanic* lies in her watery grave, the fascination grows. The more she rusts away at a depth that will always be difficult to reach – so deep, fortunately, that we'll never be able to bring her to the surface – the more permanent her

The *Titanic* only really became immortal after she sank

place in man's collective memory. *Plus belle que la beauté est la ruine de la beauté.*

The mixture of sadness, excitement and curiosity at seeing a temple in ruins, an old woman with only one tooth, a fragment of a lost poem, leads to an emotion that has no function on other occasions. It is known as the "ruins feeling". It can, indisputably, be located in the abdominal region, and has the same salutary effect as staring at an open fire or listening to the splashing of waves. I first became aware of it during a school outing, when we visited the prototype of ruins: the remains of a medieval castle. The dungeons used to be here, the guide told us, and that's where the boiling oil was thrown from above. It was a stage whose play could easily be imagined. Like all young boys in all centuries before and after me, I went and stood in an opening between the battlements and poised my imaginary arrow on the imaginary string of my imaginary bow. *Ivanhoe* came alive before my eyes:

> With that he bent his good bow, and sent a shaft right through the breast of one of the men-at-arms, who [. . .] was loosening a fragment from one of the battlements to precipitate on the heads of Cedric and the Black Knight. A second soldier caught from the hands of the dying man the iron crow with which he heaved at and had loosened

the stone pinnacle, when, receiving an arrow through his headpiece, he dropped from the battlements into the moat a dead man. The men-at-arms were daunted, for no armour seemed proof against the shot of this tremendous archer.

Maarten van Heemskerck (1498–1574), *Ruins of Septimius Severus' Septizodium*

I eagerly fulfilled the duty of every visitor to historic sites and scratched my name into the age-old walls, thus becoming – as I understood even then – part of history. It was a magical activity, much more elevated than the graffiti-spraying through which today's youth strive, in vain, to become part of the present. Here, the seeds were sown for what is regrettably so lacking in today's society: a sense of history. Nothing is more suited to making history come alive than ruins. Ruins are time on hold.

My hunger for ruins was more than satiated during a boat trip to Germany. Trips down the Rhine are trips past ruins. For a romantic young boy, the robber-baron castles perched high up on the rocks were the climaxes of a week full of stodgy gravy, cheap wine and schmaltzy songs. From the boat, or as a reward after a steep climb, the ruins gave our trip a false sense of purpose. This is how the tourist industry began: not with trips to the sea or to beaches but to ruins. Before the advent of mass tourism, the education of every well-to-do West European was rounded off with a Grand Tour to the roots of the civilization he was in the process of making his own. Today, in the Greek vases and Roman busts found in English and French country estates, you can still see how, during the seventeenth century, the souvenir trade was already busy trying to empty tourists' pockets.

As is often the case, the affluent followed in the footsteps of the artists who'd gone before them. Soon after the Italians rediscovered classical antiquity, artists from the North journeyed to Rome and Naples. Their curiosity whetted by the prints they'd seen in the rare books on the new trend, they went to view antiquity with their own eyes and to rediscover the rediscovered in person. Jan Gossaert, who set off to see the Renaissance in 1508, was followed by a succession of Fiamminghi, painters from Flanders who settled in Italy, including Maarten van Heemskerck, Jan van Scorel and Herman Posthumus. A century later, the great master Rubens went to Italy to work for eight years. It must have been a very exciting time for painters. Overwhelmed by the southern light and inspired by all the old and new, they eagerly set to work, quickly becoming acquainted with the picturesqueness of ruins. That wasn't so difficult. A modern tourist would be envious of the state in which they found the ruins of ancient Rome: overgrown, roots thrusting their way up everywhere between broken columns, the odd head here, the odd

Herman Posthumus,
*Landscape with Antique
Ruins* (1536)

hand there. And, in the midst of this great display, we see – in a painting by
Herman Posthumus from 1536 – how people, dressed in some kind of Roman
garb, descended with torches into the cellars of Emperor Nero's Domus Aurea,
or Golden House, to admire his frescoes. Of course, then too, ruins were
already a symbol of decay that set one thinking about the finiteness of man and
his works. But they also revealed how culture, despite all war and misery,
persists. The ruins in the paintings weren't an indictment of destruction, but
a classical symbol of the harmony to be aspired to between nature and culture.
Otherwise, there's no explanation for all the collapsing walls and broken
sculptures scattered precisely throughout the idealized world of Arcadian
landscapes.

When the journey to Italy deteriorated into a touristic pastime for the better
classes, the artists didn't stay away. On the contrary. In the absence of cameras,
travellers were all too happy to take artists with them, so that they could record
in paintings and drawings all the beauty that they stumbled upon. This was
how Fragonard travelled with his patron, Bergeret de Grancourt, to Rome and
Naples in 1774, and also how, a short time later, Christoph Kniep accompanied
Goethe on his *Italienische Reise*. Eventually, through countless depictions, the
Forum Romanum, the Colosseum and the Santa Maria degli Angeli became as
well known and clichéd as the Statue of Liberty and Eiffel Tower are today.
Increasingly, country estates in the North were filled with renderings of ruin-

Jan Gossaert (called Mabuse), *The Colosseum*
(c. 1478–1532)

ous Italian landscapes, by painters
who had never set foot in Italy.

This interest in classical antiquity awakened people's interest
in their own history. Thanks to
their troubled pasts, France,
England and Germany were also
teeming with ruins. These were
to become permanent fixtures
in the Romanticism that swept
over Europe at the end of the

eighteenth century. As a result of the unification of states, a great many castles and fortifications of a great many princes, earls and barons went to ruin. Later, as the old cities began to be dismantled, a piece of ancient city-wall could always be found pining away romantically somewhere. A national trust to extract time's teeth did not exist yet. Large buildings were mostly privately owned and quickly fell into disrepair in hard times. In well-maintained estates, a growing number of paintings of less fortuitous buildings began to appear. Ruins became all the rage.

Gradually, they were so much in demand that extra ruins had to be built. During the early eighteenth century, British country-home owners dreamed of having their own Tivoli on their estates, ruined temples and all. Until well into the nineteenth century, classical, Gothic and even Chinese ruins were built in France, Germany and the Netherlands, preferably on elevated sites with a view of the estate, which was as artificially laid out as it was natural in appearance. Gothic castles with "real rust from the Barons' Wars" were ordered from specialized architects, as if it were gazebos being built. In more remote areas, hermitages were constructed, complete with hermits. Lord Bathurst's Alfred's Hall consisted of a whole series of artificial castle ruins where their owner could sink away in appropriate romantic meditations. You didn't have to build your ruins from scratch either. In 1836, Hussey moved out of his fortified house in Kent, with its tower from the original fourteenth-century castle, into a new house on the hill, so that he could watch the picturesque process of decay taking place before his very eyes. Its crumbling walls were soon overrun by plants and bathed in wild flowers, and within no time it was balm to every romantic soul.

To build ruins, if not a contradiction in terms, would at least seem to be a contradiction of natural process. Why would anyone want to create decay? Yet building for decline seems to appeal to a deep need that already manifests itself at an early age. As a child, almost everyone has built a sand castle close to the water line when the tide was out and enjoyed waiting for the tide to come in. Wave by wave the castle was destroyed, first timidly, bit by bit, then, once the tide gained momentum, voraciously, so that entire walls crumbled and towers collapsed. What bliss! In reality, it's a civilized variation of the more universal drive to destroy other men's castles.

The eagerness with which the castles, fortresses and bunkers of others are destroyed is surpassed only by the persistence with which one's own castles, fortresses and bunkers are rebuilt after undergoing the same fate. Often the mess is caused by the original inhabitants, either as part of a scorched-earth policy or as a result of reclaiming lost possessions. In any event, it's the destiny of every castle, fortress and bunker to go to ruin. Those who carry shields ask for them to be impaled.

The most striking thing about military ruins is the benevolent silence that reigns over them. Few places are as peaceful as battlefields after the battle, when the winners and losers have long since disappeared. Is this also true of modern wars? You wouldn't say so if you look at pictures of the aftermath of the bombing of Rotterdam or the Trümmerfrauen (Ruins Women) desperately digging around in the ruins of Cologne. Yet there were photographers who registered the destruction of their city purely for aesthetic reasons. In Germany, in 1945–46, Herbert List recorded the beauty of the desolation that was once Munich: perfectly composed and lit, the remains of

Trümmerfrauen (Ruins Women)
in Cologne, 1954

the buildings and sculptures appear frighteningly impressive. In Cologne, in 1949, Chargesheimer envisaged a book called Form und Urform (Type and Archetype). Only a few years after the night of 31 May 1942, during which 1,000 bombs were dropped on that city, killing 20,000 people and destroying 70 per cent of all the houses, he wanted to make a deluxe publication of it. Not as an indictment against war, like Gesang im Feuerofen (Singing in the Ovens of Fire), the book of photographs that Herman Claasen had secretly shot during the war, but to enjoy it. "Ancient ruins are beautiful," wrote Weiss-Margis in the accompanying text. "We always knew that, didn't we? So why are we so shocked by our own ruins? In ancient ruins, it's also only the artistic form, born of the spirit, that's been preserved and that we see, isn't it? Does it really matter today that their buildings were destroyed in centuries and ours in only minutes?" Naturally, the book was never published; the survivors in Cologne had more important things on their minds after the war than the glorification of the misery in which they found themselves. The rubble was cleared away as quickly as possible, the destruction disappeared under the countless new buildings of the *Wirtschaftswunder*; the stains were erased. With the notorious exception, that is, of the Kaiser Wilhelm Memorial Church in Berlin, which, in one of the most pretentious locations in that capital of capitalism, tries to keep the memory alive – and only partly succeeds. In the early 1980s, the church was covered in scaffolding because the ruins were in need of repair. Seldom has so little respect been shown for the symbolic value of ruins as in the restoration of this church. Even more shocking, though, in terms of its significance as the Great Warning, is the discovery that every tourist makes if he's receptive to it: the restoration is beautiful. More beautiful, at any rate, than all the dissonant symphonies of glass, steel and concrete that surround it. In that sense, the Kaiser Wilhelm Church meets a requirement that Hitler's architect, Albert Speer, set for buildings. According to his "Ruin Value Theory", buildings should be constructed so that they, like Greek and Roman temples, would also one day be impressive as ruins. Seldom has a theory been tested so quickly in its implications as this one.

Ruins are the result of poverty, war and destruction; the work of evil. Is that why it's wrong to enjoy them? Is this the darker side of our nature seeking an escape? How decadent do you have to be to be able to enjoy destruction? As decadent as the rulers who have the destruction on their conscience? "A heartless pastime" is what Henry James called his own study of ruins. "And", he admitted, "there's something perverse in the pleasure." If war is related to sadism, ought ruin-lovers to be accused of masochism? Possibly. But, then,

Children love haunted castles

it would be the kind of masochism a child would understand. Children love haunted castles, witches who eat children and bat soup. If it doesn't have something to do with original sin, this delight in what frightens, this nostalgic longing for precisely what you want to escape, must at least have something to do with *Weltschmerz*, the world-weariness that can already affect a baby in its cradle. To satisfy this need, though, it isn't necessary to create havoc in reality. Our minds are easily appeased with sagas and legends, or their modern variations – films, books and television. Literature is imbued with horror; war films are still a favourite genre; touristic guides about the massacres of the First World War have turned out to be bestsellers 80 years after the event. Shakespeare, Greek dramas, Marlowe and James Bond provide us with plenty of dungeons, torture, bats, revenge and destruction. Even in the Bible, it regularly recurs that "one stone stays not upon another", and Job as well as Sodom and Gomorrah are eventually destroyed. "Human construction and destruction seem to have originated on the same day," as the Dutch author Anne Pingeot puts it. In Biblical stories of destruction, "to discover beauty, one [has] to wait for the refinement of decadence, for a time of decay". The question is whether that time hasn't already arrived.

Of course, there's no harm in a certain degree of open-mindedness when it comes to enjoying decay. One Dutch writer discovered this when, at the age of 52, he went to the zoo with a group of little girls. Much of the pleasure of going to zoos comes from the nineteenth-century feel of all the little temples and other romantic shelters. But the longing for decay can also affect the animals themselves. In the ape-house, the writer tried to interest two of the girls, "both four years old, in a young, lively, little monkey that was jumping all over the place, but they wouldn't budge from where they stood – in front of an enormous, flabby, Buddha-like orang-utan, who lay on the ground, like a heap of old rags, looking at them. Occasionally, despite himself, he would move an ear or a hand, at which they would shriek with delight: 'He's moving!' I told them there was a cute little monkey around the corner that moved infinitely more than that big bag of flab, but they weren't the slightest bit interested. 'We can do what we like!' replied one of the girls crossly."

The writer found it "puzzling" that the children could see anything worth looking at in that old ape. But ruin-lovers know better. They can enjoy themselves to their heart's content in old zoos. While the whole institution of the zoo is a reference to a lost era – to the days when upright citizens used to take their afternoon strolls there with the animals as a décor – something still remains of that past: dilapidated oriental ponds, chamois poised on scrappy fragments of rock, busts of exemplary founders and, naturally, the grounds themselves, which, as a rule, have successfully resisted all modern notions of gardening. The animals, too, are clearly shadows of their former selves. Deprived, as all prisoners are, of their natural environment, they display vestiges of their original behaviour, along with behaviour patterns typical of the incarcerated, such as pacing back and forth, head-shaking and glazed stares. Some animals are even the last remaining specimens of their species. That's the wonderful thing about zoos: extinction takes place right in front of you. What the management prefers to call the breeding of endangered species might as well be called controlled extinction – with a large dose of lively, screaming children thrown in for contrast.

It's this mixture of old stones and young life, weariness and vitality, nature and culture, which gives ruins their souls. There's almost no description of ruins that doesn't talk extensively of the life they house. A brief reference to the disorderly state of the walls and arches is invariably followed by a summing up of the shrubs sprouting up everywhere, the mosses, the profusion of flowers and the procession of animals: owls, foxes, bats, frogs, wood lice.

Yellow corydalis (*Corydalis lutea*)

Ruins are ecosystems, like ditches or heaths. In the same way as heath-dwellers have features in common, so plants or animals have to meet certain criteria before they're allowed to inhabit ruins. Biologists would emphasize the special roots of the wall vegetation, the way the nests are anchored and the difficulties of seed dissemination, but romantics prefer to see the animals in a sinister light: hollow-eyed night revellers, not flying but flapping. Ruin-dwellers always live in darkness, not because they can't bear the sight of day, but because it suits our romantic notions better. Also, as one journalist put it, "perfect ruins are inhabited by creatures on whom the original inhabitants preyed: from earwigs and wood lice to owls, bats and rabbits."

Biologists only became interested in ruins many years after the Fiamminghi and Jacob van Ruisdael had recorded in great detail the growth of the flora on the weather-beaten walls. The first articles on wall flora are reported to have been published in France as late as 1861. In those days botanists were usually apothecaries. They recognized many medicinal herbs from their own pharmacopoeia in the plants on the battered castle walls. Once imported from faraway countries into the gardens of cloisters and castles, apparently they had escaped and come to settle on the walls. But ornamental plants, such as the ivy-leaved toadflax, also came to the city and embankment walls via castle ruins and pleasure gardens. In 1732, it was found on almost all the city walls along the Rhine. There lay an abundance of old towns and castles, like stepping stones for the dissemination of wall plants. Yellow corydalis and pellitory-of-the-wall made grateful use of them as well. Many of our plants are addicted to ruins. If Ludlow Castle were to disappear tomorrow, we'd probably lose such rare wallflowers as the *Cheiranthus cheiri*. By the same token, ruins can't survive without their plants. They're both their inhabitants and their creators.

Aside from the kind of plants you might give on Mother's Day, ruins abound with flora from the seamy side: fungi, mosses, algae. Garden-lovers are glad to see the back of them. We have a more ambiguous attitude towards the largest fungi – toadstools. Toadstools are the plants of the mind. Medicine men, high priests and shamans, as well as hippies and even young estate agents, are familiar with their hallucinogenic properties, but it was centuries ago that people decided whether the spirits they awakened were good or evil. Names like dead man's fingers, witch's butter, death cap, destroying angel, devil's bolete and stinkhorn made toadstools suspect. That is, until about a century-and-a-half ago. During the Victorian era, a veritable mania swept over all of Europe, a mania for collecting and cultivating ferns, mosses, toadstools and other lower plants, known collectively as cryptogams. Biologically that name doesn't mean much anymore, but it lives on in the name of the comic figure, Monsieur Cryptogame, better known in English as Bachelor Butterfly, the hero of Rodolphe Töpffer's early nineteenth-century satiric stories. Monsieur Cryptogame may be better remembered as a butterfly-catcher, but in fact he was more of a toadstool-picker.

The term "cryptogam" comes from the Swedish naturalist, Carl Linnaeus, who helped give plants their scientific names during the eighteenth century. With this term, he was referring to the last of the 24 classes he had identified. Like monogams, which do it with one, and polygams, which do it with many, cryptogams do it *crypto*-ly or secretly. Cryptogams were the plants in which Linnaeus had difficulty finding the sexual organs. And that was precisely the criterion he had used to identify the other 23 classes. Botanizing consists largely of counting pistils and stamens. Today, innocent schoolchildren are still required to peer at the sexual organs of flowers. No doubt the sexual import of what the flowers are doing escapes them, but Linnaeus, who referred to flowers as "bridal beds", knew better, and his enemies knew best of all. Johann Siegesbeck, a scholar from St Petersburg, wanted nothing to do with Linnaeus' "filthy" system, full of such "shameful whoredom", which the Creator would never have allowed in the plant kingdom. Today, we know Siegesbeck only from that measly, unsightly little plant that Linnaeus named *Siegesbeckia*.

Compared with the oversexed, sperm-shooting higher plants, with their exhibitionistic sexual organs, cryptogams are chaste spinsters. That was handy for the Victorians; the predilection for ruins, Romanticism, toadstools and chastity all fitted together perfectly. Early in the twentieth century, lovers of the lower plants felt obliged to defend themselves in the Netherlands too. "They often have difficulty", wrote Dutch naturalist Catherina Cool in her famous *Paddenstoelenboek* (Book of Toadstools), "justifying to themselves why they actually choose toadstools instead of the fresh, blossoming summer children."

> It is true:
> There is something mysterious about toadstools; it is mostly their quick blooming – ostensibly out of nothing, usually with almost no visible root, out of rotting leaves or worm-eaten stumps – that has something impenetrable and mysterious about it. And when such a thing – like the widespread fly agaric – also has very poisonous or madness-inducing properties, then it is clearly the work of the Devil! Yes, they are the children of darkness, no less, "who live only from stealing and thieving, begetting themselves from rotten waste and the refuse of other plants or their remains".

"The Dutch are mycophobic," cultural philosopher Lemaire maintains to this day. "By contrast, the real mycophiles should be sought in Eastern Europe and especially Siberia." Lemaire stumbled upon the Juchen, the Yukaghirs and the Tungus as the classic fly agaric eaters. In the Netherlands, people limit themselves to toadstools and the occasional chanterelle or beefsteak mushroom. That the Dutch still love toadstools is thanks to both the toadstool exhibitions that used to be organized at schools when they were young and the father of nature conservation in that country, Jac. P. Thijsse, who smuggled toadstools, as the only lower plants, into his beloved children's stories:

Toadstools are the plants of
the mind

Actually, they're all beautiful, the little ones and the big ones. Unfortunately, many people still think of death and decay, and destruction and ghoulishness, but none of these are applicable. Today, everyone should know enough about toadstools to be no longer afraid of them.

An earthstar in itself looks mysterious and significant. This is all the more true where a fairy ring of earthstars is concerned. You should go and sit in the middle of such a ring sometime, patiently, for half an hour, and even if you are the most sober kind of person, you will begin to understand how some people can be so deeply affected by the mysterious poetic charm of these plants.

And so toadstools were domesticated. It was no longer necessary to be afraid of them. But Thijsse wasn't so dumb as to rob them of all their mystery. In a similar way, toads were eventually liberated from being ogres into being creatures that you would help across the street. Poets have sung their praises, celebrating their eyes and especially their little hands, but without losing sight of the Queen Victoria-ness, the wartyness, the pug-facedness, or their sudden appearances out of dark, dank corners. To be attractive you also have to be a bit repulsive. There always has to be something ambiguous in the object of one's love, whether it be a toad or a person.

Bats only became emancipated recently. Feared for centuries as evil's henchmen and fliers-into-old-women's-hair, they have suddenly become as popular as birds. People who once set out on a beautiful day to spy on the spotted redshank now sneak around

For centuries bats were feared as evil's henchmen

Birds' wings have been reserved for angels, bats' pinions for devils

in the dead of night with bat detectors. The irreconcilable appears to have been reconciled. Since time immemorial, birds' wings have been reserved for angels, and bats' pinions for devils. Jeroen Bosch depicted fallen angels as bats as early as 1500, but their pinions are even older. Dragons already used them for flying around to announce their evil intentions. Farmers nailed bats to their sheds to fend off evil. It's a good thing there were no bat detectors in those days: thanks to this method of tracking them down, many more bats have been found in Western Europe than was ever thought possible. Most churches are filled with more bats than believers. Bats love ruins; they live in old clock towers and hollow trees. In the Netherlands, their favourite haunt is the St Pietersberg, just outside of the southern city of Maastricht. For centuries this beautiful hill has been abused – as a stone quarry, a hideout for smugglers, a sanctuary for fossils and, during the past 70 years, as a source of almost ready-made cement. The country's most beautiful "mountain" has been destroyed in the name of postwar reconstruction. Through the extraction of limestone, the southern system of underground corridors has disappeared forever. But there's still a maze of 20,000 subterranean passages, with a total length of more than 300 kilometres. Here, the silence of darkness is broken only by the squealing of bats; 75 per cent of all Dutch bat species live there. The mountain has given birth to a bat.

Slowly slowly at first, then gradually
  more boldly
His pinion parachute took shape,
The membranes, the baleen, the nails,
The wonderful nails. He, Lucifer's bastard,
  Hell's crown prince.

And suddenly he rose, flew, ruled over
My room with his clean, black innocence,
The innocence of his evil. I loved him.

BERTUS AAFJES

Where there are feelings, there are smells. Lovers smell of roses, cosiness smells of food, fear smells of sweat. But which smells belong to the "ruins feeling"? What do ruins smell of? Damp earth and moss, fungi, cellars and heavily urinated-on lampposts. It's an ambivalent smell: it reminds you of the past, but not without the terrifying smell of rotting. What's rotten reeks; it's a warning not to touch, and certainly not to eat, it. Of all the omnivores, man, in particular, has to be careful about what he eats. Something that smells bad is therefore much more unpleasant than a good-smelling thing is pleasant. Occasionally, it's more exciting to eat something that stinks: to help ourselves over the revulsion we call it a delicacy. The more expensive and rotten, the more delicate the delicacy. And the more decadent. Seen in this light, surströmming must be the most delicate of delicacies. By eating surströmming, the Swedish keep the rest of the world at arm's length. Surströmming starts out as herring in the Baltic Sea, in the northern part of the Bothnian Gulf. When it is preserved, ingredients are added that counter the preservation process: the ingredients make the fish rot. It's ready to eat when the tin starts bulging. You have to be Swedish to dare to open it, because a nauseating, acrid stench hits you in the face. The Swedish take the smell in their stride. After all, it's a delicacy.

The Romans already had a predilection for rotten fish. They added *garum* to almost everything. This salty fish sauce has earned the Roman kitchen the reputation, among modern gastronomes, of culinary torture. *Garum* is crushed, rotten fish. As befits a delicacy, there were many different recipes, each with fervent supporters and detractors. But, basically, it was a question of leaving the salted entrails of sardines, or whitebait, or whatever was available, to bake in the sun until they were fermented through and through. Sometimes it took a few months, but then they could be made into something really special. In ancient times, the Jews had a special kosher version, *garum castimoniarum*, in which only fish with scales were used. The most expensive form was *garum sociorum*. This was made from fish that had been soaked in *garum*. Whoever abhors the thought has nothing to fear when travelling to Rome today; *garum* is no longer on the menu, except in Thai and Vietnamese restaurants. There, the food is full of rotten-fish paste – dishes such as nam-pla or nuos-nam. If you don't know any better, they taste wonderful.

The Dutch aren't even aware that their national fish is rotten. At school, they learn that Willem Beukelszoon van Biervliet developed a better method of preserving herring. In fact, the partial gutting he proposed served not to preserve the herring better, but to let it rot. To conserve food, just add salt. That was already widely known in the days before Willem Beukelszoon. But not too much of it, because the herring has to rot a little. How else does one eat raw fish? During the gutting process, the fisherman chops up the digestive glands. Since fish have no saliva, these glands are in their intestines. The digestive juices are released and, uninhibited by the salt, start to digest the fish meat; a gutted herring eats itself. The trick now is to polish off the little creature before it polishes itself off. In reality, new Dutch herring is in the same state as the half-digested fish pap that fish-eating birds regurgitate for their young.

Of the flat-fish, sole is an internationally prized seafood. Everyone knows this, but not everyone knows why. It's because sole rots better. Freshly caught sole is no more delicate than plaice. But after both fish start rotting, you begin to taste the difference. Plaice gets worse with time, sole better. It's the rotting process that makes sole so good; it's only at its best after about two or three days. So, avoid places that claim to sell fresh sole; either the chef is stupid, the owner simple or the fish foul.

Fortunately, fish rot quickly. This is for the same reason that Whitby is quiet in winter: the water's cold. While the enzymes of a warm-blooded creature are accustomed to working at 37 degrees Celsius and quickly perish in the fridge, the enzymes of cold-blooded fish feel at home in a few degrees above zero. Despite the fridge, in dead fish the enzymes continue digesting to their heart's content, something you can easily see and smell.

Fish were created on the fourth day, mammals on the fifth. That's why meat lasts longer in the fridge. But not everyone wants well-preserved meat. Hunters are as mad as the Swedish: they're mad about rotten hare. It must be a noble animal. To elevate it to this status, they leave it outside the fridge for a week. "In roast game", according to a brief euphemistic description in a hunters' manual, "the decomposition of certain proteins takes place, which

Hunters love decay as well as death

makes the meat more tender and gives it a more typical game flavour." Hunters love decay as well as death.

Almost everyone who's not a hunter or a vegetarian prefers to go to the butcher for fresh meat. But they've gone to the wrong place: butchers don't sell fresh meat. That's obvious to anyone who's ever read a detective story. If the meat in the window were fresh, it would be stiff, or – brrrrr – become stiff in the shopping bag on the way home. Meat is a piece of dead body and dead bodies stiffen. As a coroner would know, a detective wouldn't have a chance of solving his mystery at a butcher's: the meat there has already been dead for so long that the suspect has long since taken flight. In human bodies the first signs of rigor mortis become visible four to seven hours after death – in the face. A few hours later all the muscles in the body have become stiff, a condition that disappears after about twelve hours, and then only gradually. It takes about 36 hours before the whole body goes limp again. If your butcher were to sell human meat, you'd be able to see, from its degree of limpness, that it was at least 36 hours old. Other mammals have exactly the same kind of muscles as we do. What butchers sell isn't meat but carrion. Your first-class butcher is a carrion-dealer.

It's not so bad: it just means you're a carrion-eater, not a meat-eater. Meat-eaters, like lions, kill their prey single-handedly and eat them fresh, while the blood is still warm. We carrion-eaters prefer to wait until the carrion has a little more taste and smell, and then dish it up with lukewarm gravy. Today, the time it takes for meat to be hung is determined by the slaughterhouse, and by the factories where the meat is converted into super-market portions. Once home, the meat goes right into the fridge or freezer. Before fridges appeared on the scene, the hanging used to be completed at home. My grandmother, for example, was an expert carrion-eater. Before she ate the butcher's carrion, she kept it in a meat safe. This was a cube-shaped lattice-work structure covered with mosquito netting to form a three-dimensional insect screen. During the summer the bluebottles soon appeared on the netting, licking their lips, like tomcats gazing at saucy titmice on the other side of the windowpane. The smell attracted them from far and wide and also told them which phase of decay the meat was in.

In pre-fridge days, grandmothers had many other methods for allowing rotting to reach a certain stage – thus far and no farther. Herrings were pickled, ham smoked, beans dried. The herrings were so salty you couldn't eat them, the ham was carcinogenic because of the smoke and the beans were inedible if they weren't soaked first. But there was no alternative – until the fridge arrived. People continued to buy salted herring, though, and to serve smoked ham and soak beans after they'd been dried, often in factories designed especially for that purpose. The most expensive are smoked or pickled according to traditional methods; the thought of handwork and folklore enhances the taste of the past.

Cheese tastes strongly of the old days. But then, cheese is no more than an age-old method of preserving milk. Milk occupies up to ten times less space in the form of cheese and can be kept a hundred times longer. Because bacteria in the milk have converted the sugars into acids, most organisms don't like cheese anymore. The main exception is us. But we're certainly not the only ones: numerous insects, fungi and acid-loving bacteria share our tastes. Over the centuries we've become so accustomed to the sourness that we even pretend cheese tastes better because of it. If you can't beat them, join them! True-blue gourmets think old cheese should be so old that the maggots crawl out of it, or better still, jump out of it. Cheese skippers, the maggots of the cheese fly, jump across the kitchen table in leaps of 10 to 15 centimetres. That is, if they don't stay behind in the tunnels and get eaten together with the cheese. Gourmets needn't feel guilty about this: there was a time when cheese maggots ate us; they're corpse-eaters. In maggots' eyes, the discovery of very old cheese must have been a dream come true: at long last, food without awful hair or hard bones, and yet with the taste that reminded them so intensely of their lost paradise. The taste of the past.

Maggots tend to be too much of a good thing for the modern cheese-lover. What he wants is holes. Most of the holes are made by bacteria, which shamelessly expel carbon dioxide as they digest their food. Fortunately, you can hardly smell it, but that's not very important. Real cheese-eaters like nothing better than disregarding the most obvious sign of inedibility – stench. In essence, the same bacterial processes take place in the damp seclusion of cheese cups as in the moist lukewarm folds of our skin. Hence the association between cheese and feet. Where feet are concerned, people are totally put off by the stench, but stench associated with cheese attracts people, at least the French – so strongly that they recommend their Camembert in the words of their poet Léon-Paul Fargue: *Les pieds de Dieu.*

Where there's French cheese, there's generally French wine: rotten grape juice together with rotten milk. Wine is nothing more than the rotten juice of raisins. Half of the juice gets drunk by fungi – yeasts – which, out of gratitude, urinate alcohol into the bottle until it's full again. At a certain moment, the alcohol content is so high that the yeasts succumb to their own waste. Despite this, the taste of wine is thought to improve with time. Waste products from the juice and fungi form complex compounds and guarantee a stronger taste and more headaches. The age of the wine is indicated on the label, and the extent to which you should appreciate it, on the price tag.

In countries without a wine culture, this tomfoolery has been adopted by whisky and cognac manufacturers. In the Netherlands even gin is upgraded – by calling it "malt spirits" and adapting the price tag to reflect the number of years it's been in contact with oak. The basis for this trick, though, was already laid when the distinction was made between old and young gin. The drinker of old gin pays considerably more for a little sugar, a little food colouring and one extra juniper berry, even if his drink isn't a day older than the young gin in the bottle next to it.

If you can't tell how old gin is, and if cheese is being aged more quickly all the time, how can you be sure they're as old as they're claimed to be? And why should you care? This is asking for the sake of asking. "Tasty" and "rotten" are subjective concepts strongly influenced by things that have nothing to do with them: in this case, language. Cheese is never "rotten", it's "ripe"; rotten grape juice is called "wine from a good year"; and cognac derives a venerability from its age that's enough to make old people jealous. It is culture that decides where the boundary is between rotten and ripe. The

British pronounce fish rotten before the Swedish, cheese rotten before the French and bananas rotten before the Jamaicans. Whereas people from the tropics think bananas should have a healthy black colour, an Englishman won't tolerate a single brown spot. To satisfy both populations, the plantations deliver two kinds of bananas: one so thoroughly black you could make purée out of it just by touching it and the other so unripe you can scarcely cut it with a machete.

Originally a fruit-eater, man attaches great importance to the ripeness of his fruit. Very fresh fruit isn't edible; it gives you a stomach ache or is simply sour. But unlike other inedible food, fruit doesn't have to be cooked to become edible; it prepares itself. Enzymes ensure that the starch is converted into sugars and the colour of the fruit tells you how far the process has advanced. Like all apes and monkeys, people are experts at judging this. They like doing it, too, as you can see at markets.

This background explains much of the similarity between greengrocers and antique shops. In both the wares gleam brightly; in both appearance is what counts. Fruit must shine – it's a sign the contents have been properly sealed by a layer of wax – and it must have the right shade of colour to be ripe. With furniture the subtle differences between polished walnut wood and lacquered pine wood would be missed by the untrained eye, but as with fruit, these differences can tell you something about the age. If a wood surface becomes uglier with time, it's called rust or mildew; if it becomes more handsome, it's called patina. How surfaces develop depends on the material involved: brick and slate improve with age, so that stones from antiquity are extremely valuable; aluminium and steel become worse for wear, which is why modern buildings become less valuable with time and older ones more. In the old days builders were aware of the effect of time on material; today they no longer build for the future. Even artists once appreciated how time could perfect the imperfections of their buildings and paintings. Now, no one has the patience for this anymore and art is intentionally made to look old. It's business, after all.

People pursue antiques as enthusiastically as they discard old junk.

Television programmes and magazines try to teach them the difference between junk and antiques, art and kitsch. The subject is inexhaustible, because the boundary between the two is continually shifting. What was junk yesterday is antique today. From pot shards to trashy icons, everything becomes antique if it's old enough. When all is said and done, you pay for time. Time is money.

You don't have to be old yourself to appreciate the beauty of ageing. Young people have been wearing jeans for donkey's years. Originally cherished for their indestructibility, the only thing that matters now is their destructibility. A new pair of jeans is supposed to look like an old pair. You can easily distinguish old denim from new through the colour – indigo – which fades beautifully with wear. Impatient as young people are, they buy pre-faded jeans. Until recently, the manufacturer had only two methods for meeting this demand for poor quality: first, the diligently applied dye was made to fade with bleach, then the jeans were roughed up with pumice stones. Today, there are also enzymatic methods, which spare the environment and leave the warp intact. The question is whether everyone appreciates the latter: every so often, fashion dictates that holey jeans are hip. Old people need have no worries about young people knowing how to ruin things in a romantic way.

Old men's faces are a source of inexhaustible interest

Ripe or rotten, it's still a question of quantity. Very rotten is rotten, a bit rotten is ripe. Overripe is also all right, as long as it's not overabundant. Ruins are beautiful, as long as it's not the whole city that's reduced to them. A toothless old man adds charm to your Greek holiday, but 30 minutes in an old people's home is more than most can bear. Sitting on a terrace, we can spend hours looking at swarthy old faces imagining all that life has held for them. You'd have to be made of stone not to be moved by Willem de Mérode's poem "De grootvader" ("The Grandfather"):

> The last glimmer of life hung over him
> In farewells and having to renounce
> The small joys that give life sweetness,
> Something broken in voice and gesture.
>
> And when he calmly tarried down the sunny path
> We were greatly moved.
> Were God's angels abducting him?
> We saw that he had almost no shadow.

Old people inspire a mixture of fear and compassion. We think it picturesque when they linger a little towards the end, but mostly we're glad we're not them. It's easy to attract the public to a film about Notre-Dame Cathedral if there's a hunchback in it, but a pirate without a wooden leg isn't worth tuppence and *Great Expectations* would never have been so great without little Pip. It's all wonderful, as long as it's not us who's the hunchback, one-legged or orphaned. At school we couldn't hear enough stories about lepers, whose fingers would fall off if you shook their hands, or who had fewer toes every day. Rabies was another favourite. Tramps, war orphans, little match girls, how wonderful misery can be – as long as it doesn't affect us. And as long as there aren't too many poor souls.

No one has described the pleasure of other man's miseries better than Dickens. The best way to enjoy a hired tramp – who for a few coins endures cold and hunger beneath your window – is yourself to be sitting in a well-heated room with a steaming plate of food in front of you. That's the basis of one of the most romantic of human activities: charity. Throwing crumbs to the poor from your own cup of plenty. To enjoy misery, you simply have to have a minimum of comfort yourself. That's why decadence always costs money, as national museums show, with their indoor souvenir shops and cafés and their outdoor taxi stands. In their gleaming and brightly lit galleries hang paintings of poor shepherds, castles lying in ruins and destitute families – Octave Tassaert's *Une famille malheureuse* – all well-cleaned and restored. And in galleries of modern art, with their white-plastered walls, Gilbert and George, believers in "the artist as art", put themselves on display stark naked; and neatly dressed men and women wearing glasses that are somewhat too gaudy admire Damien Hirst's bisected calves or his fellow artists' half-rotten pigs. Occasionally, there's a cry of indignation, but in essence, decadent art differs little from modern nature conservation. Everywhere in areas protected by The National Trust, dead and fallen trees are left untouched, to allow nature to run its course. Sometimes trees are even cut down to help it on its way. Rotting is part of nature, The National Trust tells us. But those homes in the middle of those estates full of fallen trees continue to be restored to the hilt: spic 'n' span, not a speck of moss anywhere in the gutters, fungi exterminated, ferns extracted – as if old wall plants don't ever get hungry. Fungi and mosses and ferns are never left to eat to their heart's content. If this isn't even allowed in natural habitats, where will it be allowed? Where in Western Europe can you still find genuine ruins, beautiful devastation, a building falling apart in its own time – overgrown with vines, undermined by ants, under a blanket of pigeon droppings, the ideal home for spiders and lizards?

As a boy in Amsterdam, I felt at home with ruins. After the war, many neighbourhoods were just rubble, the chemical industry wasn't yet obsessed with eradicating fungi, dockyards still had empty sheds and the acute housing shortage kept the demolishers at bay. Where today the Dutch National Bank stands in Amsterdam, proudly erect, the gallery of the *Paleis voor Volksvlijt* (the

School children find
stories about lepers
are endlessly
fascinating

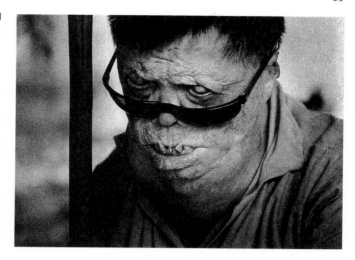

now burned-down Dutch version of the Crystal Palace) once stood, gloriously disintegrating – until the postwar reconstruction of the country. Who, nowadays, comes across a house declared unfit for habitation? All the crumbs that have been left behind by time's teeth are removed by dustbusters. Why, one asks? When the nave of Utrecht Cathedral blew down in 1674, the ruins were left lying around until 1826. For a century-and-a-half, little boys were left to play in them and old men to fantasize about decay. Today, a windy square lies between the tower and the transept. You can see where it once used to be beautiful from the contours of the pavement.

Ruins are not only beautiful, we also need them, writes J. B. Jackson in *The Necessity for Ruins*. You need a mess to be able to build a new future on an idealized past, when everything was better:

> . . . history [is] not [a] continuity but [. . .] a dramatic discontinuity, a kind of cosmic drama. First there is that golden age, the time of harmonious beginnings. Then ensues a period when the old days are forgotten and the golden age falls into neglect. Finally comes a time when we rediscover and seek to restore the world around us to something like its former beauty.

But there has to be an interval of neglect, there has to be discontinuity; it is religiously and artistically essential. That is what I mean when I refer to the necessity for ruins: ruins provide the incentive for restoration, and for a return to origins [...]. Many of us know the joy and excitement not so much of creating the new as of redeeming what has been neglected [...]. The old farmhouse has to decay before we can restore it and lead an alternative lifestyle in the country; the landscape has to be plundered and stripped before we can restore the natural ecosystem; the neighbourhood has to be a slum before we can rediscover it and gentrify it.

As children, we knew all too well when the golden age was: "before the war". Quality was still produced then, there were still genuine Dutch wares, strawberries still had a flavour. At school the Golden Age turned out to be the

Utrecht Cathedral after the
storm of 1674

seventeenth century, but even as a teenager I never quite managed to rid myself of a longing for the nineteenth century. The golden age isn't the problem though; the future too shines, sometimes a little too brightly, all around us, even though it has only just become the present. But where did all those indispensable ruins go? Everything that was old has gone or been restored. How can a child grow up to be well balanced in such a world? Where is the middle-aged biologist supposed to go to daydream? What's the point for an old person, when everything around him is young? How to live in a world that denies decay?

Give us back our ruins! Throw a few crumbs to the fungi and the beetles – a little villa here, a little warehouse there, an abandoned waterworks site over there – something the creatures can really get their teeth into. A waste of old buildings? It doesn't have to be old buildings; nature loves new buildings too. Just make a few holes in the gutters or rain pipes and within no time they'll be the ideal mouthful, thanks to the sour urine of moisture-loving micro-organisms. As well as a Monuments List of old buildings earmarked for restoration, there should be a Ruins List of new buildings earmarked for ruin. I have a few suggestions, if anyone's interested.

3

CRUMBLING ON A GRAND SCALE

"Blackie's going grey," a woman friend once said to me. Seldom has ageing been announced so succinctly. It moved me. Cats, too, turned out to be mortal. The world's most lovesick animal, the best thing creation ever produced, could also go grey and lame and suffer increasingly from chilblains.

Everything that's whole has to break down: my wristwatch, the Taj Mahal, Queen Victoria, the Severn Suspension Bridge, even Blackie. You can only be whole for a short time. Wholeness is an exceptional state, as fragile as a house of cards – or me. Once I was dead, now I'm here for a short time, and before long I'll be dead again, but then forever. That's the way it works, for everything. First there's nothing, then there's something, then there's nothing again. Nothing lasts. And strangely enough, we can almost live with it. But who, or what, keeps breaking everything? It's enough to make one angry. If only we could get our hands on the bastard! What kind of a Creator is it who makes the most beautiful creations on earth, only to turn around and let His other, equally beautiful, creations usher them off to their maker? Is the Creator a Vandal? Might it not have been better if He'd left it to us?

The best things ever made by man himself are the Seven Wonders of the World. It's still a mystery to human minds that human hands were able to create such things. For all their computers and cranes, modern engineers would be hard-put to build anything so magnificent. Here, we rivalled the gods. And? Did they last? Of the Seven Wonders, six now lie in ruins. The Hanging Gardens of Babylon – once the pride of Queen Semiramis, planted in thick layers of soil on the palace roofs, irrigated with pumps from the river

All that remains of the Seven Wonders of the World are the Egyptian pyramids

far below – are deserted, desolate, dust. After 1,500 years, the Pharos of Alexandria collapsed into the sea during an earthquake. And antiquity's statue of liberty, the Colossus of Rhodes, succumbed to enemy attacks. The statue of Zeus, draped in gold, was also destroyed by plunderers. Today, you have to go to the British Museum to see the last remnants of the Temple of Artemis. There, you can admire the sculptures Sir Charles Newton had shipped from Ephesus to London in 1856. The remains in present-day Turkey have long since had nothing to do with the original World Wonder; it was almost completely destroyed by fire in 356 BC. If there is one World Wonder that has been a total disappointment, it's the Mausoleum, whose only real task was to defy time. That beautiful tomb, surrounded by 36 Ionic columns, covered by a pyramid of steps and crowned with a triumphal chariot, was built to perpetuate the memory of Mausolos, King of Caria, after his death in 353 BC. An earthquake destroyed it in the thirteenth century. The Knights Hospitallers built their castle on its ruins. All that remains of the Seven Wonders are the Egyptian pyramids. Yet they too have failed in their mission. The sphinxes have let the treasures they were there to guard slip through their paws, and the Pharaohs, who sought everlasting life, have wound up on display for budget tourists. Five thousand years, that's how long eternity lasts.

But the Pharaohs almost succeeded. Some arranged for themselves and their treasures to be buried so ingeniously that local tomb desecrators couldn't get to them for centuries. It's thanks to those preventive measures that there was so much for European tomb desecrators to steal during the past 200 years. Never before has so much of an ancient culture ended up in museums as that of the Egyptians. And never again will so much of an ancient culture be lost, because once something has been excavated, time can really get its teeth into it. While mummies change hands – from dealers to swindlers and from exhibition to exhibition – time's appetite remains insatiable. Gold gets remelted, tourists are cheated, vases fall irreparably into pieces or are irreparably restored. Art can be ancient or modern but it isn't everlasting. "Art perishes," wrote art historian Gary Schwartz, "if not now, then later. More art is lost than saved in every generation. Destruction, not survival, is the norm." In 1971, Edward B. Garrison earned the scorn of his colleagues when he remarked that at least "70 or 80 per cent of all the paintings made in Italy during the twelfth and thirteenth centuries must be considered lost." Spurred on by his critics, he later adjusted that figure to 99 per cent. Another art historian calculated that this figure would be just 1 per cent less for German medieval altar panels. That can't only have been the result of a lack of respect for art because, according to another historian, not more than one-tenth of all the paintings that must have existed at the beginning of the Dutch Golden Age, the heyday of painting in the Netherlands, were still around at the end of the seventeenth century.

What one hopes is that the scanty remains represent the *crème* of culture. After all that sifting through by all those generations before us, surely only the best remains. If only that were true! In reality, the most beautiful things are usually destroyed, so that the ugliest survive. Attacks on neighbouring populations involved not only burning their villages and raping their women but also destroying their art treasures, which, given the evil of the enemy, could then be nothing but *entartet*, or degenerate. Worse than neighbouring populations are neighbouring generations. Any normal child hates what he inherits from his parents. Even though Mum and Dad mean well, they can still have bad taste. One generation's culture gets thrown out by the next. But worst of all

Museums discard as well as save, sometimes
even a stuffed orang-utan must go

are your own people and your own generation, and worse than the worst of all is you. Just think of all the things you've thrown out during your lifetime as undesirable reminders of the past you! You don't need a disaster to be stripped clean. "Moving three times", wrote Benjamin Franklin in 1758, "is as bad as a fire." But what else can you do? You can't keep everything; you'd expire under the weight of your own possessions. That's how curators and archivists see it too. Their main tasks aren't to save and store but to discard and destroy. At the Public Records Office, 90 per cent of what comes in is destroyed, so that the other 10 per cent can be saved. There, it's mostly paper, mountains of it; in the case of museums, it's sculptures, old cars, stuffed orang-utans, paintings and, of course, the hobbyhorses of previous directors. Gary Schwartz isn't very impressed by the gatekeepers who separate the wheat from the chaff:

Not only do artists and art historians lend oblivion a helping hand when they perform their daily duties, they're also perfectly capable of participating in less lofty, art-destroying practices: iconoclasm, vandalism, political repression, making military targets of enemy art treasures, cultural genocide, illegal excavations, removal from public supervision, wilful neglect and euthanasia for difficult pieces of property. Add this up – together with the effect of natural disasters, the billion-dollar market for stolen and smuggled art that often disappears forever, the accelerated deterioration of precisely those works we value most and thus that we ship around and exhibit most – and very little is left of the flattering image we have of ourselves as the guardians of the past. We're depleting our cultural reserves as quickly as our natural resources.

Nostra culpa. Man destroys his own creations. As children, sooner or later we break the toys we love most. The only mitigating circumstance is that even without our help, few of our creations would survive. After gold rushes, whole towns fall apart as a matter of course, and temples of not-even-very-old civilizations are swallowed up by surrounding jungles. Wherever maintenance is neglected, even if only briefly, blades of grass force their way up through the asphalt, cellars flood, thick slabs of concrete sink into the earth. Culture can only be maintained by constantly keeping nature at bay. Stagnation is decline. How is it that nature always knows how to get the upper hand? As is often the case with difficult questions, the answer must be sought among the exceptions. Just take a look at a packet of Camels and you'll know why the pyramids have lasted so unnaturally long: they're in the desert. There's no water there – that's the definition of a desert. So, where there is no desert, there is water. It comes down from the sky and the mountains thanks to gravity, creeps up capillary-like through the smallest cracks and seeps behind everything. If need be, it makes its way by force. In its transition to ice, water suddenly takes up so much more space that rocks split and pipes burst. Once the ice has forged a path, the water can start on its own devastating work. In moderate climates this happens every spring; in many other places it happens every morning, when frosty nights turn into scorching days. The warmer it gets, the better water can do its work, because it absorbs solid materials more readily at higher temperatures. It's not just a coincidence that you can clean things effectively with water; more than half of all existing elements are soluble in it. Some substances are so eager to be dissolved that they slurp moisture up right out of the air. That's why kitchen salt becomes moist. Wherever water flows, it takes the most water-loving substances with it, until the substratum is so undermined that it starts to crumble. The dissolved and disintegrated substances can then be swept away. But substances disappear from sight when water stays still too. Once dissolved, many compounds become unstable. In solution, kitchen salt separates spontaneously into its composite elements: sodium and chlorine. Sodium is positively charged, chlorine negatively. Normally, the two would have a strong affinity for each other, but this isn't the case in water. The electrical properties of water reduce their force of attraction to 1 per cent of normal. As a result, the sodium and chlorine particles float about aimlessly, or they tarry for a while near the oppositely charged part of the water molecule. If, in addition to the kitchen salt, another substance dissolves in the water – and that's always the case – the different particles readily bump into each other. This is how new combinations, new substances, are formed. Without the cooperation of water, the original particles would never have been able to muster up the necessary licentiousness; water offers them ample opportunity to change partners. This is why salt has such a destructive effect on cars. A car can be parked in a mountain of salt without suffering any damage at all as long as water doesn't show up to activate the salt.

Whole buildings and whole mountain ranges dissolve in water like sugar in tea. But a cup of tea can't absorb a whole sugar cube. The tea would be cold before it finally reached the heart of the cube. First, the sugar disintegrates into granules, then it seems to disappear like snow in summer. Initially, water has little effect on buildings and mountains, but the more cracks and crevices appear, the more the crumbling advances, and the better water can get to where it needs to be. The surface contact between water and stone increases quickly every time something breaks away, whereby the crumbling process accelerates, the water gains more access, and so on. By the time a mountain disintegrates into rocks, and the rocks have become pebbles, the pebbles sand and the sand mud, the shared surface of the particles has approached the infinite. Now the water can almost lick up the molecules, something facilitated by the fact that all the silt eventually gets washed out to sea. By this time, the particles can no longer be saved. The longevity of stone depends mostly on how well it manages to stay away from water in its early days. Sensible mountain ranges protect themselves against erosion by fostering luxuriant growth; in wet climates, buildings are as good as their gutters. But life isn't easy; Leon Battista Alberti, the "redis-coverer" of the building rules of classical antiquity, complained about rain even in the fifteenth century: "Rain is always prepared to wreak mischief, and never fails to exploit even the least opening to do some harm: by its subtlety it infiltrates, by softening it corrupts, and by its persistence it under-mines the whole strength of the building, until it eventually brings ruin and destruction on the entire work." So, you would say, build as far away from water as possible. That was no problem where the pyramids were concerned, but cities are intentionally built close to rivers or at the intersection of water-ways. The reason Cracow's historical inner city is in such a bad state today has less to do with war than with attempts to prevent it. The city was purposely built between the tributaries of the Vistula to deter attackers. As a result, the ground is swampy and the buildings never really dry out. This is exacerbated by the fact that the city is in a bowl and so is not well ventilated. In mild weather, the moisture penetrates deep into the walls of the Royal Palace; when it freezes again, extensive damage is caused. And Poland, thanks to communism, managed to escape the insulation wave that buried houses in the rich West under a mountain of environmental regulations! In the West, water's not only kept out; it's also kept in. In the absence of natural ventilation, the water from our kitchens, bathrooms and lungs hangs around as vapour before condensing in the wood joints, behind layers of paint and between floors. It thus reaches places it would never have been able to access as a liquid.

Water is so busy dissolving substances that it's seldom found in pure form. A rain drop even absorbs substances as it travels through the air: carbon dioxide, for example. This is the gas that gives natural water its fresh taste. Truly pure water has no taste at all – which is all the more reason to sell mineral water instead of tap water, and to enliven its taste with an unnatural dose of carbon dioxide. But, today, there are more substances in the air than just natural carbon dioxide, and they dissolve too. This is why rain is becoming more rotten all the time, more acidic. Environmentalists raised the alarm in the 1970s. Trees couldn't stand it, they said. Mass "forest deaths" were predicted. If that prediction had come true, Europe would now be as bald as the bottoms in advertisements for women's underwear. So it wasn't that bad after all. It was mostly the environmentalists who turned sour. As for the affected forests, it's mostly our own fault. Those forests that are still left in Europe have to grow on the poorest soil, and where deciduous forests should be, coniferous trees are doing their best to survive with roots that grow much too close to the surface. An etching by Albrecht Altdorfer (1480–1538) already reveals the effects this can have. It depicts two spruce trees clearly showing the symptoms that today we would attribute to acid rain: desiccation, drooping branches of the second order and loss of needles. Given its title, *Landscape with Two Young Spruce*, it would be difficult to diagnose the syndrome as ageing. Today, we blame the deterioration of European forests on a complex of factors, only one of which is the degree of rain acidity.

If anything, culture seems to have fared worse than nature. What first seemed to threaten man himself now mostly seems to attack his images. As gentle rain turns into drops of acid, facial expressions on carvings in ancient cities become contorted grimaces – lepers, noses rotting away, ears falling off. Old sculptures that have survived for thousands of years succumb to the exhaust fumes produced by the driving style so popular in precisely those countries where most of the sculptures are found. Photographs from the last few decades reveal how sculptures have lost their contours, like well-licked lollies. Given the importance of the tourist industry, the first preventive measures have been taken. The caryatids of the Erechtheum in Athens and the four giant bronze horses on the façade of St Mark's Basilica in Venice have all been replaced by replicas. The bronze statue of Marcus Aurelius has been moved from the Piazza del Campidoglio to the Capitoline Museum in Rome, thus killing two birds with one stone: the statue is safe and the tourists have to pay to see it. The realization that the best things our civilization has produced are only safe behind museum walls is enough to make you a cultural pessimist. Stones that defied countless earlier centuries can suddenly no longer withstand our century. But is this because of our century, or because of our stones?

In the days when the hole in the ozone layer was still called acid rain, photographs of St John's Cathedral, in the southern Dutch city of Hertogenbosch, invariably appeared in government brochures as an example of the damage caused by acid rain. The faces of "the Holy and consorts" are eaten away most on the west side, where the rain can get at them best. Deformed, they look up at the heaven that has brought them so little, or down at the cars whose exhaust fumes have given dechristianization a helping hand. The restoration of St John's seems endless: pinnacles have been removed and reinstalled, arches repaired, paintings exposed. Statues, too, have been taken in hand and, where necessary, completely rehewn. Not one of the figures on the huge flying buttresses flanking the cathedral dates from the Middle Ages anymore. Even so, it's not the medieval stones that have suffered most. Acid rain has dug most deeply into the Saint Joire, a limestone used for the church's restoration a hundred years ago. Lime dissolves much more readily in acid than the components of the original Bentheimer and Gobertange stones. As a result, many of the statues remade during the previous restoration have had to be recarved all over again. I found it fascinating to watch the sculptors at work. After all, something of great biological significance was taking place. By creating one statue in the image of another,

Stone is no more immune to decay than the faces it depicts

they were not only endowing the statues with the ability to contract diseases, such as leprosy, but also giving them the most glorious of gifts: the ability to reproduce. Now that's bringing stone to life!

But what's the point of making images out of stone if stone is more transitory than the figures it depicts? This question is even more relevant in the case of Cracow, where materials that are too soft have been used since day one: sandstone from the Carpathians, and porous pinczów from an area northwest of the city. Both types of stone can be shaped easily – by man and by acid rain. Sandstone is nothing more than glued-together sand, and that glue unsticks in acid rain, after which the stone disintegrates into powder. Rust and deposits adhere well to the rough exterior, so that the pores become clogged and the moisture can't escape. This moisture allows the calcium in the lime to unite with the sulphur in the rain, resulting in calcium sulphate. When the moisture evaporates, the salt crystallizes, expanding with such force that the stone falls apart. According to Stepien, the curator of the Royal Palace, this effect has been known for centuries: "Old books describe how armies secretly left salt in the walls when they were forced to retreat."

But destructive as water itself is, one of its two components – oxygen – causes even greater devastation. No element is as effective in demolishing houses, trees, and whole cities and forests. After all, fire is nothing more than the process of uniting with oxygen. But the process doesn't have to manifest itself in the form of fire. In fact, it's the gentle smouldering – of underground peatmoor fires, food being digested or whole libraries turning yellow – that causes the worst damage. And even though water is fire's worst enemy, in the case of gradual decay, water and oxygen are partners in crime.

Together, water and oxygen are consuming the world. It takes them ten years to work their way through a one-centimetre-thick plate of steel. Assisted by water, oxygen combines with iron to form a compound as delicate as it is complex. The iron crumbles. Initially, the first thin layer of rust protects the virgin steel beneath it, but it's too delicate to do that for long. Copper's rust does a better job; green copper oxide keeps oxygen out for a long time. But the best protection is still paint or varnish, except that they, too, then come into the firing line. Paint and varnish have to be renewed regularly because oxygen knows how to deal with them as well – this time assisted by light. Light is simply a form of energy. Pigments absorb that energy until it literally makes them burst open; their molecules fall apart, whereby the colours change. White people turn brown on account of it. A short time later (it's not always an undivided pleasure) they're just as white as they were to start with. But changes in the colour of paint are not reversible. To prevent their paintings, Gobelin tapestries, stuffed birds or books from fading even more, museums keep out the most dangerous light – ultraviolet – with coloured windows, and reduce the artificial light to a minimum. Outside, though, the window frames and gutters continue to peel as of old. There, water, oxygen and light have a free hand, and only the pipefitter and painter can help.

Wind and water also find their way into the museum director, the museum director's wife and you and me. We breathe them in deeply about 10 or 15 times a minute. This is how wind and water enter our lungs, where the oxygen is removed. This extremely aggressive substance gets distributed to every corner of our body via the blood. In such a moist environment, the organs have no choice but to rust. There's not one scientific reason why this word should only be used to describe the oxidation of metals. If we inhaled pure oxygen, our lungs would rust so quickly we'd be dead within a few days. If the atmosphere contained only oxygen, we wouldn't even exist. Forests and moors and everything that grows in and on them would spontaneously combust. So it's a good thing the oxygen in the air is diluted with four times as much nitrogen. But this doesn't mean it's been deactivated. The greatest danger of all probably comes from oxygen in the form of free radicals, once called "molecules from hell" by the *Chicago Tribune*. A free radical is a molecule that misses an electron. To compensate for this, it steals an electron from another substance, damaging it in the process. The body defends itself with substances that spirit away the free radicals, but the system is not foolproof. The DNA of every cell is attacked thousands of times every day. Most of the damage gets repaired, but scars remain. The DNA ages as a result and, with it, the whole cell; man rusts away from the inside.

Pilgrims queue for holy water at Lourdes

While ships sink in it, mountains are washed away by it and works of art are damaged by it, diehards hope to benefit from it (with a bit of help from God, of course). An average-sized public utilities company would be proud of the amount of holy water that changes hands in Lourdes: in little Maria glasses, jerry cans, wine bottles – everywhere, you see people who can barely walk dragging such things around with them. The only thing missing is a fire hose. The holy water may bring everlasting life within reach, but whether the transitory will be prolonged as a result remains to be seen. According to the British weekly the *Catholic Herald*, holy spring water is a hotbed of contagion; patients with a weak constitution can die from it. On several occasions, staff in two British hospitals have noticed a deterioration in the condition of patients who had been sprinkled by next-of-kin with holy water from distant holy places. Microbiologist Karen Allen found countless dangerous pathogens in it. In the Royal Preston Hospital in Lancashire, nurses immediately confiscate any holy water that is brought in, so that it can be sterilized. Only then is it returned to the visitors.

If the heavens gave us sterilized water only, buildings and mountains would last longer. Raindrops contain invaders that should frighten us much more than the Normans, the Huns and the Barbarians. With an unimaginable number of pseudopodia, grappling hooks and probosces (or beaks), bacteria, algae and fungi settle on the highest skyscrapers and mountain peaks. They form the vanguard. Thanks to microscopes, we know that they contain trillions of teeth, in billions of jaws – or sometimes no teeth at all. If it weren't for the fact that our ears are used to it, the sound of all that chewing and sucking – amplified 100,000 million times – of all that rotting going on all around us would clearly be audible as a symphony of decay. A piece of beef on the kitchen counter would take only 24 hours to sound like Tchaikovsky's 1812 *Overture*. The Huns were paragons compared with these little plunderers. Most of them dislike new buildings. There's little for them to latch on to and the fresh mortar tastes like uncooked beans. But something's being done about that. The advance guard of micro-organisms opens, so to speak, the back door, so that the wind and rain can do their work.

Studies conducted on fragments of stone by Patrick Jacobs, a geologist from Ghent, revealed that marble, mortar and sandstone are scarcely affected by the Belgian climate – until micro-organisms are released on them. "That noticeably affected the stones. Some pieces completely fell apart. Mosses and fungi are the worst. They develop threads that can penetrate deep into the stone. A microscopic fracture or hole is all it takes. If that happens, something has to be done." But what? Often, the first thing one reaches for is a hose; the wall is sandblasted or sprayed clean. Professor Marcel De Cleene once watched how St Michael's Bridge in Ghent was sprayed clean with water from the Leie River. He knew it was a big mistake, because the pressure meant the algae and bacteria would be sprayed deep into the stone. Working closely with wind and water, they prepare the masonry for less menial organisms. At first, the wind doesn't help much, because after a downpour it quickly blows the wall dry. This is why these little pioneers can only live in places where it leaks; life is good under a leaking gutter or next to a broken rain pipe. But everything changes after the arrival of the mosses. Their roots work their way into the stone with their acid juices. Remove a tuft of *Grimmia* moss, and you'll see the grooves of its roots clearly marked in the sandstone. The traces are even more visible in polished marble. Mosses can grow into layers ten centimetres thick on slate or thatched roofs, and there's no question they damage the roof with the water they retain. That moisture, in turn, attracts higher plants. If a roof or wall is favourably located in the sun, decay quickly becomes a delight to the eye, in the form of greater celandine, yellow corydalis and pellitory-of-the-wall. The seeds of these plants are transported by ants who, as a reward, develop special lipid deposits. Birds spread seeds with their droppings, and wind, too, as a redistributor, again proves itself decay's faithful accomplice.

It's no coincidence that the first thing you think of when you hear the word "rotting" is fungi. Where fungi are found, humidity is usually so high that it's almost impossible to stop the decay. No fungi can survive in less than 20 per cent humidity, and to settle somewhere they need much more than that. This is usually where there's a leak or a lack of ventilation. The latter is less noticeable than a leak. In my recently renovated house, fungi ate right through a new layer of joists in less than two years because I forgot to open the cellar window. Like creatures from a different, plant-ruled planet who are searching for extra accommodation, they had arranged themselves in thick cushions around the joists, unaware of the heavy bookcases above them. What surprised me most was that it wasn't even that damp in the cellar. Later I learned that house fungi only need moisture when they're young. Once they have their feet firmly on the ground, they produce their own water supply from the cellulose in the cells of the wood. My wood. During the war, house fungi flourished in London, where water from the fire extinguishers after the bombings provided the necessary temporary humidity. The fungi were then able to move on from those ruins to penetrate other, dry houses. Using their special runners, fungi conquer walls and any other obstacles that get in their way. Sometimes they produce so much water from all they've eaten that it trickles out of them. This is why this particular species has been given the Latin name *lacrymans*: the "tearful one".

Outdoors, you only find house fungi in woodsheds or on telegraph poles. But, as might be expected, the types of wood fungi that help eat our houses still originate from forests and moors. This explains their haste. Wood fungi have to eat trees at the same rate as the trees grow. The fungi aren't fussy about whether it's timber or tree trunks, as long as the wood is dead and thus easily digestible. But a forest, too, consists largely of dead wood, not only in the form of fallen trunks with picturesque toadstools growing out of them, but also in the trunks of living trees. While dead branches fall off outside, the heartwood dies inside. It no longer has to grow or transport nutrients; the bark and sapwood do that. The only thing heartwood has to be is solid. It's very good at that, which is why it's used for carpentry – unless fungi get to it before the carpenters do. This is also why there are hollow trees. Very old oaks are completely hollow inside; by that time you can't tell how old they are anymore because most of the annual rings have disappeared. But this doesn't mean the fungi limit themselves to the heartwood. The young, newly formed wood is also a close-knit community of living and dead elements. Every so often, one out of every ten pine trees in our forests has to be chopped down because the tissues have become clogged with the *Sphaeropsis sapinea* fungus. The wood develops black stripes, and within a few weeks the tree expires. Another infamous conifer-killer is *Heterobasidion annosum* – an assassin, even according to Jac. P. Thijsse:

> He begins his attack in the dark – underground – at the neck, where the trunk of the young pines becomes the root. There the spores push their mycelium threads inwards; the mycelium then spreads best in the layer between the bark and the wood, and in the bark itself. The roots can thus no longer get food from the green branches, because it moves downwards through the bark. They starve and rot in the process, and the tree dies.

Flat crusts develop at the feet of languishing young trees in the autumn. The honey fungus (*Armillaria mellea*) has a much more handsome toadstool, which terrifies forest rangers but is a delight to enthusiasts; at night the toadstools

Honey fungus (*Armillaria mellea*)

sometimes emit a ghostly, phosphorescent light. As if to comfort the forest rangers, there are also fungi which are themselves plagued by other fungi. The parasitic bolete (*Boletus parasiticus*), for example, grows exclusively on the common earthball (*Scleroderma citrinum*).

Toadstools shoot up with proverbial speed as a result of the way they grow. Both the mycelium and the toadstool consist not of cells like ours but of long, thin threads. Threads are the ideal way to reach as far as possible using as little material as possible. Because of their structure, toadstools never grow as tall as oak trees, but the mycelium can spread over an area as large as a whole country. The only prerequisite is a thin film of water between the walls of each hyphal thread and its environment to allow the transfer of food, because this most hungry of time's ravagers has no teeth. Fungi can cooperate perfectly with organisms that do have teeth, however, for example, with the Mexican spider *Mallos gragalis*. This spider doesn't act the way a spider should. First, it lives with thousands of its kind in an enormous web. Secondly, it doesn't clean up the remains of the insects that it sucks out. Here and there half-eaten flies are intentionally spun into the web. Strangely enough, the flies don't give off an acrid odour; instead they smell rather nice – of yeast. The whole spider colony spreads an odour that attracts flies; this keeps the spiders happy. As they're feeding on the flies, the spiders inject an anti-bacterial antibiotic into them. This gives the fungi, which are always present in flies, the opportunity to eat the rest of the fly, which would normally be consumed by bacteria. The proof of this is when the yeasts attract more flies than the spiders can handle. At such moments, the bacteria take over and the colony begins to smell so strongly of rotting that no fly would come within 100 kilometres of it.

We know about different smells from our experiences in the kitchen. If you leave meat outside the fridge for too long, bacteria start to make it smell. Fruit, on the other hand, rots away rather attractively in the fruit bowl. It's too dry for most bacteria but juicy enough for moulds. It's a feast for the eyes to watch how the fruit of a higher plant is taken over by a lower form of life. Although there's no toadstools on the surface of rotten oranges, there is reproduction. The white or black powder left at the end is spores, which are always eager to polish off oranges before the world expires under their weight. We don't think moulds improve the taste of oranges, but many creatures depend on them to predigest their food; moulds are their cooks. Wood is inedible for most creatures until it's been marinated, so to speak, by moulds. This is why it's precisely newly restored churches and city halls that are so eagerly attacked by the death-watch beetle (*Xestobium rufovillosum*). This beetle begins its life in the oak beams in the form of a rather large woodworm. During restoration, the roof structure lies open for some time. The moisture that enters the beams during this period can no longer be blown dry by the wind once the roof is closed, so that the moulds are free to take their place at the dinner table, followed by the beetles. To prevent this happening, so many holes would have to be bored and filled with poison that the beams would be undermined. *Goede raad is duur.* Good advice is hardest to find when you need it most.

If no solution for this problem is found, thousands of old buildings will collapse. But the problem pales in the face of what's happening to the millions of books that are on the shared menu of moulds and beetles. Since publishers changed from rag paper to paper made from wood fibres, woodworms have been re-educated into bookworms. Keratin-lovers, such as moths and fur beetles, have joined them at the table. Insects with fancy Latin names like *Trogium pulsatorium*, *Dermestes lardarius* and *Attagenus unicolor*, sink their equally Latinized mouth-parts into expensive folio volumes and rare India-paper editions. What we see as a library, they see as a restaurant. One insect bores its way greedily through the paper. Another nibbles exclusively on the glue so that the book falls apart. Yet another, the bacon beetle, does both – if its faeces are white, it ate paper; if they're brown, it ate glue and leather – but it also likes mattresses and chilli peppers. The firebrat (*Thermobia domestica*) only removes the sheen from the paper. After devouring a book, the death-watch beetle will happily go on to eat the bookshelf too – as if paper made from wood fibres wasn't already having a hard enough time! At one point, alum resin was added to paper to make it easier to write on. This leads to a chemical reaction that produces sulphuric acid. The use of gallnut ink only reinforces this effect. The result is that nineteenth-century documents are often in a worse state than those on paper from earlier centuries, when paper was made from finely ground rags mixed with calcareous water.

It's easy to blame our great-grandfathers. But since their day, it has gone from bad to worse. It's as if our information carriers are becoming increasingly transitory. It's true that paper from the nineteenth century crumbles in one's hands, but celluloid films from the early twentieth century were so inflammable that they were stored in bunkers for fear of explosions. But at least these films lasted nearly 100 years. The videotapes that record life only a decade ago are scarcely worth looking at. During the twenty-first century, all information will be recorded digitally, but that, too, has a great disadvantage: it will only be decipherable with the help of equipment that becomes outdated more quickly than swimwear in St-Tropez. The result is that a text in Babylonian cuneiform script will preserve better than any of tonight's television programmes.

Documents on paper made in the nineteenth century are often in a worse state than those on a paper stock of an earlier century

Given the impermanence of both their medium and their message, newspapers are a symbol of transitoriness. A mayfly made of paper. What you eagerly reach for one day is only good enough – so the cliché goes – to wrap fish in the next. As if fish were any less transitory! Decay is as inherent to newsprint as the food in your shopping bag is populated by an army of micro-organisms that fancy themselves in heaven. Fruit and vegetables, with their thick cell walls, last longer than meat or fish, but once cooked, the cells of plants and animals are totally helpless; their walls have been demolished and their contents are there for the taking. This is precisely why we cook. But we're not in the kitchen just on our own behalf. Bacteria and moulds, too lazy to cook for themselves, share our food. The trick is to finish off everything on your plate before the microbes do – immediately after the food has been cooked, because cooking sterilizes it. Problems only arise when you save cooked food. Bacteria love leftovers. English microbiologist John Postgate studied what happens to leftover stew when it is put on the kitchen counter at eight o'clock after clearing the dinner table. By that time it has been thumbed, sneezed and coughed on and bombarded with bacteria; as the plates were cleaned, for example, eight staphylococci moved from your thumb to the leftovers. Thanks to cooking, the starch in the potatoes, the carotene in the baby carrots and the proteins in the meat have already been pre-digested. Moreover, the leftovers are pleasantly warm. Thus encouraged, the staphylococci begin reproducing while they eat. By nine o'clock, there are 20 participants; by ten o'clock 40 have joined the party; and by the time the clock strikes twelve, no fewer than 160 are taking part in the orgy. By lunchtime the next day, the leftovers are swarming with a million staphylococci. They're not visible, but your stew is no longer what it once was – because whoever eats has to defecate. Although the faeces of one bacterium may not be very significant, those of a million staphylococci are. If nothing else, they produce a strange flavour.

Hygiene in the kitchen is simple: figure out what bacteria like and do the opposite. Tease them, torment them to death. They like water? Dry the food. To keep our food out of their mouths, one just has to remove the water in it. This is how Eskimos store cod – as stockfish – and Indians store buffalo meat – as pemmican. Dried fish or dried meat can be kept for years. So can dried humans.

Since 1609, four corpses have been lying in a vault under the chancel of a little church in the Friesian village of Wieuwerd, refusing to make their final

Above and opposite: The Wieuwerd mummies

exit. Thanks to the dry draught, the bodies have withstood the ravages of time rather well. There's not much left inside them: at one place in the body farthest to the right, which was claimed by tuberculosis, it is possible to look right through it, via the abdomen and back, but aside from that, the outside is more or less intact. What's left resembles a dead insect. In insects, too, the soft insides decay first, so that the little creatures hang around on earth for a while like empty drums, until specialized teams of demolition organisms finish them off too.

As if the four corpses aren't macabre enough, a few birds have been suspended above them. They're dead too, of course, have been for years, put there once to see if the secret powers of the vault also worked on birds: a canary, a starling, another canary, a bedraggled parrot and a rooster, hanging upside down in a row, on cords, wares of a poulterer gone mad. The starling looks a little cross-eyed. The birds live on in their underground coop, as dour and distorted as the mummies in the coffins beneath them. Every year thousands of tourists shuffle past the mummies, which grant them a glimpse of a different world: the underworld. That's where the Wieuwerd mummies really belong, yet they're still close enough to the living to function as a sort of *trait-d'union* between the two worlds. It evokes a sensation from long ago as if, having never been abroad, one is standing on the cliffs of Dover and imagining oneself in another country. Or, as if, much later – but still many years ago – one is standing in West Berlin peering out over a 100-metre-long wall into that strange country hundreds-of-thousands-of-square-kilometres big; a glimpse. The realm of the mummies is, if possible, even more absurd.

In their world, the little eater-uppers are on strike. Something must be wrong; the mummies are grinning suspiciously. A glimpse may be exciting, but a head-on view might be too much of a good thing.

Assen, a city not far from Wieuwerd, is home to the remains of twelve people who became stuck in the peat hundreds of years ago. Their facial expressions have been frighteningly well preserved. Here, nature has used a different method of conservation: pickling. The acids of the high peat bogs tanned the skin but dissolved the bones. This is in contrast to what the substances in low peat bogs do. They are much less acidic, so that the skin is lost and the bones are left behind. Between 1791 and 1951, 48 bog bodies were found in the Netherlands, and given such charming names as the Yde Girl, the Weerdinge Couple and the Zweeloo Woman. Hundreds of such bodies have been found in the rest of northwestern Europe. Many of them were intentionally killed and sacrificed. In Denmark, the Tollund Man was found with the same noose around his neck that was used to hang him. After

more than 2,000 years, his serene expression still sets one thinking. When peat cutting stops altogether, many more bog mummies will never be found; the British discovery at Lindow in 1984 was probably one of the last. This made the finding of an ice-man in a glacier at Tirol in 1991 all the more exciting. Not only his leather jacket but also the hay in his shoes was still intact – after 4,000 years. Here, nature used the method of conservation that's most common today: cooling. This doesn't kill bacteria, but it does severely impede their reproduction. The colder the better. There's no reproduction at all in a deep freeze, but the rate of reproduction is also very low in a fridge. This isn't so strange. After all, how eager would you be to reproduce

Tollund Man

at a temperature that was a few degrees above zero?

Whichever way you do it, the essence of conservation is to protect death from life. To prolong death, you have to kill what's living. You can't conserve something and keep it alive at the same time. Life never lasts; you're decaying all the time. But who's to blame? Who wants to hurt us? Is it the environment? The alarm bells of the ecologists are ringing loudly, but they're out of tune – because of acid rain and because they've been rung so often. And if all those enormous bells are already rusting from acid rain, what does that say about our lungs? Or our food? All that poison in the fields, that surplus manure, that genetic manipulation – it can't be a good thing. Yet, despite our lungs being filled with dirty air and our stomachs with hamburgers, we're still growing older than our unpolluted forebears. So, that can't be the reason. But we're also still ageing. The cause isn't outside us but inside us. We're not made very well: we wear out; any fool can see that.

Put on a dark shirt in the morning and look at your shoulders in the evening. If it's not snow, it's dandruff. Snow melts, dandruff stays on your shoulders. Dandruff is skin confetti from your hair. If you have dandruff, you shed flakes. If you don't, you still shed them; you just can't see them very well. During the course of your lifetime, you lose your own weight in skin. A snake or a crab sheds its skin intermittently like a coat; we peel flake by flake, like old paint. Since it's a little oilier beneath your hair – from all the sebum secreted by all the sebaceous glands – the microscopic scales stick together to form visible flakes. In this way, during your lifetime you can see with your own eyes how you turn to dust. No wonder people don't like the sight of dandruff! But most wearing out isn't visible; it takes place inside you. That's why you feel yourself getting older. Joints become less supple, bones seem to creak. The tissues of capsules become shorter and more inflexible. Some break, the remainder adhere to form thick rigid bands. Where the ends of bones are supposed to slip easily over cartilage, there's increasing friction because of calcification. Free radicals lend decay a helping hand. Then there's that moment when you do your back in. Degenerating vertebrae, says the doctor. Is it true? Are we ready for the scrapheap after a few years, like cars?

The process of decline is more clearly – and more slowly – visible in your teeth than in your dandruff. A kangaroo grows as old as its teeth. It has to go through life with only 16 of them. The moment it can stand on its own two feet, the first four teeth are in use and the second foursome is breaking through. These last for about 20 years. By that time, the third and fourth foursomes have taken over, but after about ten more years, they too are completely worn out and kangaroos that live in the wild are condemned to starvation. Hippopotamuses also need dentures by the time they reach old age. But the best examples of the decaying process in nature are insects. Their chitin armour wears out, like jackets: first they stop shining, then the protruding parts start to break off, and finally the seams give way. The wings suffer most; after two weeks, the wings of an adult male housefly are so bedraggled that the fly can't even get airborne anymore. Sometimes a wing just breaks off, as in a poorly serviced aeroplane. Of course, the rate at which a plane falls apart depends on how often and how quickly it flies. As long as it's carefully stored in a hangar, not much can happen to it. According to German physiologist Max Rubner, much the same is true of animals. Those that lead fast lives will wear out more quickly than those that lead slow ones. It's as if every animal is born with a certain amount of energy, which it can then use up as quickly or slowly as it sees fit. The more a furnace is stoked, the sooner the coal will be finished and the sooner the fire will go out under a mountain of ashes and cinders. If body cells produce more waste products than they can deal with, they become clogged. In the end they die of their own chemical waste. You can see this in old people: the pigment in their liver spots is nothing more than chemical cell waste. Yet old people have fewer pigment cells in their skin than young people; the pigment is simply more unevenly distributed. Every summer you see proof of it. Old white skin turns brown less readily than young white skin. That's probably one of the reasons why beautifully browned skin is so popular: it's a sign of youth.

If cars wear out, together with houses and lampposts, then so do people. How could it be otherwise? But there's one big difference: people are alive, which means that what wears out gets repaired. Certainly, a person can be compared with a car, but then it must be a car full of car mechanics. German biologist August Weismann also thought in these terms when, in 1882, he attributed ageing largely to the process of wearing out: "Death", he wrote, "occurs because a depleted tissue can no longer renew itself." He assumed that animals age because of an accumulation of the damage they incur in everyday life. It was more than Weismann could have known. In addition to injuries and illnesses, the body also withstands radiation and free radicals. These deregulate not only the body but also the body's regulating system. The older the body gets, the less effective the white blood cells are in combating bacteria and viruses. Antibodies intended to deactivate foreign substances in the blood become increasingly inept at distinguishing foreign from natural substances, so that intruders are increasingly left unharmed and the body's own tissue is attacked.

In the beginning, you hardly notice that things are on the verge of getting out of hand. Until the age of 20, the new building work and repairs far outweigh the decline. From then on, you draw on your reserves. Young lungs can inhale six times more air than is needed for normal breathing. If this becomes five, you still don't notice it. The heart, too, can work much harder than is needed if the body is in a state of rest. As you get older, the reserves are depleted, but the body knows how to deal with this. It simply produces more norepinethrine, the substance that stimulates your heart to work. After the age of 40, there are no more top performances, but a wise person hardly notices it because he adapts his lifestyle accordingly. If you stop playing rugby and curb your sexual adventures a bit, you'll manage just fine.

Outwardly, by the time you're 40 a lot has changed. The connective tissue in your skin has aged: like that in your joints. But in your skin you can see it. Connective tissue has the structure of a pile of ladders: the older the pile gets, the more inter-connected the ladders become. The result is that it's increasingly difficult for them to slide past each other; the tissues become shorter and more rigid. Look at old people's skin. It still has some elasticity but it's less resilient, until eventually it sags like baggy pockets. The less pliant the skin becomes, the more wrinkles it develops. In the grooves between the wrinkles, all the elasticity has gone.

As skin wrinkles, hair becomes thinner. Hair follicles take longer breaks or stop working altogether. Moreover, each hair grows more slowly and is thinner than it used to be. A radical form of wearing out, you might say, but baldness is a clear sign of the greater complexity of ageing. If it were no more

The less pliant the skin, the more wrinkles it develops

than a wearing-out process, all men would go bald equally quickly. In reality, one already has a smooth pate as a student, while another still has a veritable mop as an emeritus. The balding process does follow a pattern though. First, slight indentations appear on the forehead, then the hair becomes thinner at the crown, and finally there's only a fringe at the back, until eventually even that disappears. The key to the process lies in the hands of the eunuchs. They never went this bald. Baldness is due to male hormones. The balder the man, the maler he is. Women usually only start going bald after the menopause, when the male hormones start playing a greater role. Their remaining hair usually turns grey because the hair follicles no longer supply pigment. This is quite unusual in the animal kingdom. Among wild animals, only African buffaloes go bald and grey; water buffaloes are grey in the prime of their life and go black as they age.

Grey hair is as poor an indicator of a person's age as baldness. It's better to look at how worn down the teeth are – as in horses, for example. But the wearing-down is no more responsible for ageing than going grey or bald. Nor is it a result of ageing. Teeth wear down from chewing, not from ageing. Even if you ate only porridge, you would still grow old, albeit with perfect teeth. Conversely, you don't grow a single day older from having your teeth filed down. The same applies to your other organs. Even if you were to avoid every abrasive activity, you still wouldn't reach the age of 125. The fact that man isn't destined to live long can't be blamed on his exterior, which takes most of the blows. It has to do with the most interior of his interior parts: the genetic material in the nucleus of each cell. This is where the instructions are given for both the big and the small service jobs. By the time you're an adult, these instructions have been duplicated many times over. Every time a cell divides, an encyclopedia of knowledge has to be copied. Errors are made in the process, as they were in cloister libraries of bygone days. After being copied a few times, the message is no longer the same; after it's been copied about ten times you wouldn't even recognize it anymore. If no corrective measures are taken, a cell can be written off within a year. The fact that this doesn't happen has to do with the repairs department. Unfortunately, the quality of the repair work steadily deteriorates. Not because good personnel has become scarce, but because, despite everything, they're increasingly poorly trained. Moreover, there are more frequent power failures. Mitochondria, little vesicles located outside the nucleus, give cells their energy. They have their own DNA, which is even more sensitive than the nucleus' DNA, even if only because damage there gets repaired less effectively.

Working at half speed and with increasingly illegible sets of instructions, one cell after another shuts down. Tissues and organs stall. Between the ages of 20 and 80, the maximum lung capacity of a person decreases on average by about 40 per cent, the maximum heartbeat by about 25 per cent. You have to stay in bed longer to get the same amount of sleep and it becomes increasingly difficult to retain your urine. Your bones become more brittle. The organs that suffer most are the ones that can't renew their cells: muscles and nerves. A muscle consists of a large number of fibres. If one fibre breaks down, it's not replaced, however vigorously you might run or skip. The only effect of exercising is that neighbouring tissues become thicker and fill the empty space. If you don't exercise, most of the lost tissue is replaced by fat. That's why the first muscles to suffer are the ones you can't exercise. Unless we do eyelid exercises, the eyelids will weaken until they droop. In the end they contain only half the original number of muscle tissues. In the eye itself, the tiny muscles that keep the pupil open become so weak that the supply of light they let in is always deficient. By that time, visibility is already a problem because the lens and cornea have become clouded. Light is dispersed in all directions except the right one. One's surroundings become vaguer and there's no other sense that can compensate for the loss. Your hearing starts to deteriorate from the age of 30 onwards. First the high tones fade. You notice it when you're in company; it becomes increasingly difficult to concentrate on one voice. Then you start avoiding the noisy pubs. Restaurants get dropped later, when your taste starts to fail. By that time, your tongue is only affected by powerful stimuli. That's why old people are as mad about sweets as children. It's not a coincidence that that's the taste reserved for the final round, when all the other taste buds have been knocked out. Other nuances are lost through the diminishing sense of smell. The number of olfactory cells, high up in the nose, decreases, and the cook receives fewer and fewer compliments. In the end, old people don't even smell their own body odours. This can be very embarrassing.

The more the senses fail, the fewer impulses the brain receives. It's as if the world is becoming smaller all the time. This may not be a bad thing, because there are fewer and fewer cells to process the information. Every day of your life you lose 50,000 of them. By the time you're old, you've lost an ounce of brain. Fortunately, it's improper to draw conclusions from this since we already know that women have one ounce less brain than men. Moreover, it's unwise. Even if brains do lose 50,000 cells a day, it would still take 500 years before they were depleted. What they miss in the form of lost cells, they compensate for by developing new connections between the remaining cells. It's not the number of telephones that counts, but the number of telephone calls. It's even quite conceivable that the brain is smart enough to discard the dumbest cells. The more dumbness you get rid of, the smarter you become. This may well be why we think wisdom comes with age.

The pitcher goes to the well so often that it finally breaks. At a certain point in time, you're old enough to die. But from what? There are plenty of options to choose from. Many people are most afraid of murder, AIDS and plane crashes. But chances are slim their fears will be realized. Two out of every five people die of poor blood circulation or cancer and most of the rest die in bed too. However commonplace dying is, though, it's still an art. It's still most successful in the theatre, where it only breaks out with a vengeance after the hero of the play has been killed with a sword between the ribs. As he dies a languishing death, he throws in a final aria; we have to endure the shrill funeral march until the bitter end. The dying man who enjoys himself most of all is the one who manages, with his last breath, to filch a promise out of his next-of-kin that they'll later live to regret. In reality no one dares hope he'll be granted as much time as this on the brink between life and death. But it's still an absurd idea that someone is alive and kicking one moment and as dead as a doornail the next. Life is too much of a good thing to be cut off with a single blow. Just try it. What's the best way to commit suicide? Where's the main switch? Hospitals are bursting at the seams with potential do-it-to-yourself murderers in whom everything is shattered except life itself; they were one floor too low. Then there are the countless desperadoes who throw themselves in front of oncoming trains, which – thanks to their poorly planned jumps – they will later spend many years being pushed in and out of, in wheelchairs. The biggest fools of all take sleeping pills, fall asleep and are indignant when they wake up again later.

It's even more of a problem to kill someone else. An animal is difficult enough, as anyone knows who's ever managed to retrieve an ill-fated bird from a cat's claws. Executioners are faced with even bigger problems. In countries where the death penalty still exists, they're still worrying about how to flick the switch as painlessly as possible. Not *en masse* – with a bomb – but individually – *à la carte*. Where to start? In the Middle Ages, an executioner began by removing an ear or a nail. Dying was supposed to take a long time and to hurt a lot. That's what the public wanted and the victim didn't, so it all worked out for the best. Today, an execution has to be as painless as possible. This dates from the French Revolution. Thanks to liberty, equality and fraternity, Joe Citizen was entitled to die in the same way as the nobleman: by the sword. His upbringing, however, had never taught him how impolite it was to lift one's head a little during the chopping. Executioners complained they couldn't do their work properly because of it. The time was ripe for mechanization: the guillotine came into fashion. Outside of France, though, in the rest of Europe, they continued to swear by hanging, strangulation, the wheel and burning. In America, jails were fitted out with gas chambers and electric chairs. The *New Scientist* compared the pros and cons of the different methods. The guillotine, indeed, turned out to be better than the noose. Together with the blood, consciousness – and with it, the undoubtedly acute pain – quickly flows out of the severed head, although witnesses claim to have seen the eyes of the head looking around reproachfully for some time afterwards. The bullet is supposed to be better still – if it's fired directly into the neck at close range. In general, though, a more circus-like ritual is opted for,

Left: Gas chamber, *c.* 1937

Right: Photograph from 1928 of an execution by electric chair. The photograph was taken with a hidden camera

in which a whole row of soldiers do their best to shoot a whole heart full of holes – from a distance. The age-old idea that the seat of the human spirit is in the heart seems to be here to stay. The gas chamber, still in use in America, allows the victim to choose between a death struggle of many seconds or one of several minutes, depending on how long he wants to hold his breath. Such an end is almost as painful as the one in the electric chair, in which the fully conscious condemned man feels himself become paralysed, suffocate and burn. Smoke rises up out of his hair, nose and ears. Fortunately, the death sentence was banned in the Netherlands more than a century ago. Now, only death itself has to be dealt with.

In Italy, buildings as well as people used to be decapitated as a form of punishment. In the Middle Ages, many towns were full of slender towers ranging from 40 to more than 100 metres high. San Gimignano had 72 of them and Pavia was known as the *civitas centum turrium*, the city of a hundred towers. The higher the tower, the higher its owner's status. If the owner fell out of favour, the tower had to be adapted. If the city condemned the owner, the tower was decapitated. To avoid damaging the neighbouring houses, the tower was buttressed by wooden posts on two or three sides. At the point where it was to be cut off, large holes were made in the walls to weaken them, after which the posts were set on fire and the tower fell in the desired direction. Justice was done, as Matthys Levy and Mario Salvadori conclude in *Why Buildings Fall Down*. It doesn't surprise them at all that buildings are treated like people. After all, they bear a strong resemblance to each other:

> A building [. . .] breathes through the mouth of its windows and the lungs of its air-conditioning system. It circulates fluids through the veins and arteries of its pipes and sends messages to all parts of its body through the nervous system of its electric wires. A building [. . .] is protected by the skin of its façade [. . .] and rests on the feet of its foundations. Like most human bodies, most buildings have full lives, and then they die.

Buildings die the way people do. They collapse. Again, it's all or nothing – within the bat of an eye. One moment you can still go inside and up the stairs to enjoy the view, the next you're standing looking down at a heap of amorphous stone. The Campanile di San Marco collapsed entirely unexpectedly in 1902. Suddenly 99 metres of stone had had enough of adorning the Venice skyline. The clock tower that stands there today is an imitation. Why did the original tower suddenly call it a day after 1,000 years? Lightning was blamed, the frequent ringing of bells, poor masonry, the foundations, but the only satisfying answer came from an ordinary Venetian citizen: "It died of old age."

If you're heading towards a destination you shouldn't be surprised if you reach it one day.

4

AS GOOD AS NEW

Why don't I have children? Because I know too much. What I'd like is a girl who's about four years old with hair down to her knees, who crawls into my lap, throws her arms round my neck and tells me I'm the sweetest daddy in the whole wide world. Give me a dozen such children. But it's out of the question for me. I know too much. I know such a child doesn't exist.

If you still want one, though, as a man you first have to find a woman, then you have to inseminate her. This is followed by nine months of anxiety. And you're never sure what will come out. One thing is certain: it will never be a four-year-old girl with hair down to her knees. At best, a bawling baby appears which will shatter your night's sleep and, if you're not careful, your wife's career. There's no turning back. So you feed it. And it stays wet, day in and day out, week after week, for four years. And then, just imagine: I'll be damned, it worked! There she is, the child of my dreams, with her long hair. She climbs onto my lap, throws her arms round my neck and says I'm the sweetest daddy in the whole wide world. I melt. And she says it again, and again, and – why not – yet again. But then it's really over. Children grow out from under your hands. Before you know it, you're stuck with a bony schoolchild and a short time later, a pimply-faced teenager who steals money from your pockets to buy joints. I

A bawling baby will shatter
your night's sleep

don't need this any more than I need a bawling baby. So no children for me.

Wise people know you have to love all four phases as well as all of the thousands of phases in between. This constantly changing scenario is what good parents enjoy most: it's the best possible proof they really did give life to that child. Every picture in the photo album is different, just the way it's supposed to be. Nature doesn't sit still for a moment.

Those who understand this least are nature's protectors. They have an ideal in their heads which they tend to nurture at all costs. The best example of the arrogance with which man tries to stem the course of nature is the oldest nature reserve in the Netherlands: the Naardermeer, a small lake about 30 kilometres from Amsterdam. It looks just the way it did when Monumentenzorg, the Dutch equivalent of The National Trust, created it. There's no clearer proof of the fact that it isn't real nature than this – if Naarder Lake had been real nature, Naarder Forest would have taken its place long ago. Shallow lakes fill up of their own accord with dead reeds, after which trees spring up out of the marsh. But Monumentenzorg doesn't allow this. So every day, men are there touching up the lake so that it looks untouched. It's supposed to look like a picture from a picture book. But nature isn't a picture. It isn't a momentary record. It's a film; a thriller.

Nothing stays the way it was. That's nature. And we don't like it. The world's already complicated enough when it's doing nothing at all. Which is why the conservationist movement is so popular. It preserves. Nature wants grass to sprout up? The conservationists release large herbivores to eat it all up. Life against life – the classic antibiotic. But whereas you can't bring life to a standstill, you can make sure it doesn't move forwards, like an exercise bicycle. Every change in nature is undone by those who protect it, after which nature tries again, its protectors intervene again, and so on, until natural cycles change into a treadmill on which you turn but also stay still. It's obvious why nature conservation is so expensive. Many human hands are needed to make nature look untouched by human hand.

You won't get very far if you take the past as the norm. Willow trees are a case in point. It's tiring to watch the willow pollarders at work in the

Netherlands. The fact that all those branches will have grown back again in a few years, the realization that pollarding or topping is about the most unnatural thing that can be done to a tree, the uncomfortable thought that pollarding would fit in quite well at a country fair celebrating old crafts, and the awareness that willows are the solitary remains of an agricultural system that has been lost forever – all this is enough to make one feel more powerless than ever, especially given how enthusiastically the volunteers do their work and how beautiful the result. Especially if seen from a distance – small men in a large landscape – it's as if the nature conservationists are trying to touch up the landscape with a fretsaw. What appears to be a war against ugliness is in fact a war against time – a war fought in the name of love. Love for everything that is beautiful and in danger of disappearing. But does someone who loves pollarding also love willows? Cutting back everything that grows reminds me of mothers who patronize their children out of love, who keep them small. And what does a good mother do who's sorry to see her child growing up? She quickly has another child and later – probably – another one. It would be better if we left all those Naarder Lakes to dry up and all those willows to grow as they please, and then dug up a new lake every so often so that the whole process of drying up could start all over again, willows and all.

Naturally, things are changing in the conservation world. New nature seems to be better than old. Dead trees have been left to rot so that new life can grow out of them; old cows are being imported into the Netherlands from the Scottish highlands so that they can watch how slowly our rivers flow through our lowlands. But the guardians of the forest who do what one's supposed to do in nature – guard – are few and far between. The advocates of new nature are more reactionary than the conservationists of old nature. They don't stop the clock; they turn it back. Restoring, not conserving; that's the new goal. This way you don't preserve what you have; you hold on to what you had. A restorer restores an area of natural beauty – or a painting or a tugboat – to the state it was once in, preferably the original one. Well-worked agricultural landscapes are dug up to become the marshes they once were, marshes are dredged into lakes, water levels are raised or lowered as if nature were a public bath. Then, once everything has been restored to the desired primal state, conserving can start all over again and nature can finally run its course – until a new fashion in nature conservation breaks out.

One can no more restore an area of natural beauty – or a painting or a tugboat – to its original state than one can turn women into the little girls they once were; conserving is ambitious enough. Yet restoration – regaining virginity – is still the aim. Just as many doctors patch up the hymens of marriageable Muslim girls, nature conservationists import endangered animal species and old-tugboat lovers take courses in riveting. Against their better judgment, they try to restore the flower to the deflowered. A restorer is the opposite of a rapist, and less successful too.

In ancient Rome only virgins were good enough for worshipping Vesta. If they acted unchastely, they were buried alive. Modern western man no longer requires that a woman be a virgin, but the longing for intactness remains. Who wants a woman with a missing part? She should be auctioned off, even if she only lacks a front tooth or a nostril. A man has to be whole too: One-Eye may be King in the Land of the Blind, but elsewhere he has little authority. Unfortunately, there's nothing more fragile than intactness. One footstep in the newly fallen snow, one scratch in the paint, one moment of thoughtlessness, and the untainted is tainted. One tiny sip and the teetotaller no longer exists; he's either "teeing" totally or not at all. Just looking at something already taints it. Nature that's been walked in, a woman who has no secrets, a painting that's already been seen by a million eyes: never again will they be as they used to be. Once filled in, the blank spaces on the map will never be blank again. What's been discovered can never be undiscovered. What remains – at best – is a sight to be seen.

Thousands of museums all over the world are full of discoveries thought to be worth looking at. Even though museum visitors aren't the first to cast eyes on them, they have one advantage over the discoverers: generally the discoveries look much more perfect now than they did when their discoverers first saw them. The broken pieces have been glued together, the copper polished, the ears stuck back on: the work of the restorer.

A restorer shows you what the past was like. But what *was* the past like, other than better? That depends on the present. The way you look is determined by your point of view, even if it's the past you're looking at. Nothing is as mutable as history. Just put an old history book next to a modern one. Or go to a western. It's not easy to decide in which decade the film is set, but it's easy to see when it was made. Cowboys with long hair date from the 1970s. If the heroines have breasts like battering rams, you've landed in the 1950s. The dumber the Indians, the older the film. And otherwise, the style of acting betrays when the film was made. Every age colours the past in its own way – often literally, too. During the eighteenth century, tones in restored paintings were lightened – intentionally or not – while during the nineteenth century, they were darkened.

We don't know what cowboys were really like because films didn't exist back then. Paintings and sculptures, on the other hand, give us eyewitness accounts of past centuries, right back to prehistoric cave paintings. That way we can look right into other times! It's wonderful – but how distorted is our view?

Although paintings outlive painters and films last longer than film stars, depictions are as mortal as what they depict. Old paint fades in light, varnish cracks, linen decays, layers of dirt hamper viewing. Cupriferous paint eats paper to bits; a leg falls off a statue. Until the mid-eighteenth century, no one worried much about these things. Broken-off noses were replaced as easily as we change flat tyres and paintings were patched up like old trousers. Sixteenth-century sculptor Benvenuto Cellini believed that the ravaged sculptures of antiquity "were screaming for his help". It was only

Every age colours the past
in its own way

during the Enlightenment that a need for authenticity emerged. Today, the code of ethics of the Dutch Association of Restorers talks of the need to "honour the integrity of the object". A restorer should interfere as little as possible, and retain as much as possible of the original materials. Interventions should be recorded so that they can be undone again later. Today, ancient objects are surrounded by men wearing coats that are just as white and creams that are just as expensive as those worn by beauticians in exclusive salons. Art has become science. Using microscopes and chromatography, the question of whether there should be any restoring done at all is posed and reposed, but invariably, restoring is decided upon. In this respect, art experts resemble tree surgeons, who are always prepared to declare sick those old trees that grow so annoyingly on the planned motorway route.

In the past, not the work of art but art itself had priority. A work of art, like a house, was regularly touched up. This was certainly the case for sculptures on display outside. By the time the interest in antiquity had flared up, the overdue repairs were so overdue that "unpainted" became the norm. A museum full of Greek and Roman statues makes a predominantly "white" impression. Renaissance artists also started using white a great deal, and later touch-ups made that white even whiter. Following the restoration in 1989 of the tomb of Ilaria del Carretto in the Lucca Cathedral, art professor James Beck complained that it looked as if it "had been cleaned with Spic 'n' Span and polished with Johnson's Wax".

As a result of the Grand Tours, the demand for Greek and Roman antiquities reached heights that have only been surpassed by today's obsession with stripped pine furniture. Obliging dealers glued together any pieces that bore a resemblance to each other, filled the open spaces with plaster and shipped the whole mishmash off to France and England for a lot of money. Urns suddenly had ears again; Venuses grew arms; for every teapot there was a lid. All antiquities collected before the Second World War have at least one loose body part. It's still difficult for us to know when the different parts of a statue were crafted, but it's no problem at all to see when they were glued together. An excellent example is Piranesi's vase in the British Museum, in

Before and after restoration. Jacopo della Quercia, Ilaria del Carretto, c. 1406–1408

which the famous Italian architect reshaped the fragments of an ancient stone vase from Hadrian's Roman villa into a new creation all his own.

The dividing line between restoring and forging is vague and shifts with time. Many restored art objects are later "unrestored" so that they can be restored again, but this time without embellishments. Pots are dissected into shards, inauthentic shards banned, and what remains – purged of tainted foreign parts – is returned to the display case as a vase – until the next restoration. But the line has shifted yet again. In Munich in the 1970s, it was quite acceptable for Thorvaldsen's nineteenth-century additions to be removed from the Greek temple sculpture from Aegina. But today, we would question whether his additions should not have been restored together with the temple itself. Opinions change quickly. In 1969, frescoes were displayed in London with great success – without the walls they had belonged to for centuries. In our time, such a thing would be unthinkable.

What has stayed at its location is Masaccio's Adam and Eve Being Banished from the Garden of Eden, in the Brancacci Chapel in Florence. However, an important part of it has disappeared: the leaves that hid Adam's genitals from pious eyes. After permission was received from the highest authorities, the leaves were removed during the 1984 restoration. They were thought to have been added to the much older fresco somewhere between 1652 and 1798. Excitedly – and in the eyes of art historians, perfectly legitimately – the public went penis-watching, but the critics had their doubts. A replica of Masaccio's banishment fresco in Rome, which was attributed to Michelangelo, had a completely different model penis. Apparently, the organ had already been

Masaccio's *Adam and Eve Being Banished from the Garden of Eden* (1425–1428), before and after the restoration of 1984–1988

removed from the public eye in Michelangelo's day, possibly even by Masaccio himself. The foliage had probably been removed during an even earlier restoration, only to be bowdlerized again later with a new leafy decoration. Who knows? And who knows what people will say years from now about our restoration frenzy? Maybe they'll just want the prettiest picture again. Maybe they'll eventually also appreciate the work of art as a result of all those centuries it has passed through, together with all those embellishments of all those other artists and especially all that dirt and fading.

Personally, I'm always disappointed when I look at a restored painting. There's something kitschy about it. What it's gained in freshness, it's lost in dignity, and no professional code in the world can undo this. When a painting is cleaned the layer of varnish that protects the actual rendering is removed. As rigorously as the layers of varnish – with all the dirt from over the years – are removed, the old paint layers are carefully left intact. Whether achieved with acetone, urine or the gentlest of soaps, the restoration washes away the age of an object that derives its value partly from that very age. However much the painting is enhanced, the feeling of direct contact with the past is lost. Edmond and Jules de Goncourt were already complaining about this in 1851, after seeing a Rubens and several other paintings for the first time since they'd been cleaned:

> It is like a piece of music from which all the half-tones have been removed: everything screams and bellows like earthenware gone mad. Robbed of their golden patina, the paintings have made us have serious doubts. Time is a great ruler; is it also possible that it's a great painter?

One art historian contrasts this approach with a caricature by William Hogarth from 1761. Hogarth didn't like the healing effect of time on art. He drew Old Father Time seated in front of a painting, smoking, his scythe piercing the canvas. The painting is shrouded in smoke from the pipe and the varnish has darkened. In the foreground a severed hand points to a pot of varnish. According to Hogarth, Time was art's enemy: "Will the water in a landscape be clearer or the heavens shine more brightly if they're brown and dark from decay?"

The cleaning controversy causes fewer problems in the East. There, it's a holy duty to keep the statues in the temples intact. As a result, the layers of paint and dirt that have overrun the original colours over the centuries are cherished. For the Japanese, even if gold glitters silently under layers of smoke and incense, this is no reason to polish it back into visibility. If you come across a restored Japanese statue gleaming brightly in an average European museum, you can be sure it's a fake.

With the protective layer alone being the subject of such heated debate now

According to William Hogarth, Time was art's enemy. Engraving from 1761

and in the past, wouldn't it simply be better to restore less frequently? Archaeologists have long since concluded that their treasures are much safer buried underground than in museums, out of reach of the eager clutches of restorers and curators, unstudied and unspoilt. It's precisely because they were unfindable for so long that they have survived at all. In one sense, it would be better if the remaining unrestored art treasures were to be buried, this time in well-sealed cellars and vaults. Until now, this gateway to eternity has only been reserved for unknown painters; at least they're left undisturbed. According to New York-based restorer Marco Grassi, to find old paintings in top condition one needs to look among the works of the smaller, lesser-known masters who escaped the nineteenth-century restoration frenzy. Their paintings are often "impeccably preserved, so we have to think something terrible started in the nineteenth century when the painter/restorer trade began". Perhaps all restorers should be banished to the same island, far away from here – and I'd be happy to see all nature conservationists sent there too – if necessary for many years, until they finally learn the most difficult thing a modern person can imagine: to keep their hands to themselves.

It's not easy to sit still when there's a lot of money at stake. Many restored pieces are worth more than the original work of art. To the owner, a work of art is often a small fortune with a frame around it, and he's not about to stand by and watch helplessly as time ravages his hand-painted money. But artists don't make their lives any easier. Increasingly, modern art is being forged out of old bicycle wheels, milk cartons and other junk designated "non-traditional materials" by the art world. Curators, art historians and physicists engage in serious discussions with restorers about whether a sponge that has been decaying since 1982 can be replaced – in this case in the name of restoring Tony Cragg's *One Space, Four Places*. This work of art consists of shampoo bottles, paper, concrete, foam rubber, tin cans and other garbage that Cragg collected along the banks of the Rhine and elsewhere. If you make art from garbage, isn't that an open invitation for the art eventually to return to its original state? This is particularly relevant to mobile art, such as *Gismo* by Jean Tinguely, an artistic snarl of bicycle wheels and other pieces of scrap metal that can be driven by a motor with a transmission belt. But if you turn it on, the art work destroys itself. Since it was acquired by the Stedelijk Museum in Amsterdam in 1974, the little hammers in the mechanism have made holes, the chassis has collapsed, the wheels have stopped turning, the belts have gone slack. In reality, every machine has these problems, whether it's art or a steam-driven pump. If it works, it wears out; if it doesn't, it's not a machine anymore. According to a former curator of the Stedelijk Museum, *Gismo* was supposed to be on at all times, so that the noise would lead you to it from wherever you were. Apparently, that curator had accepted the unacceptable – that life has to live, even if it results in death. It's a liberating thought, and one that is consistently pursued, with all the resulting consequences, by those who make works of art from ice or sand, and who then stand by, not without some satisfaction, and watch how the sun or water reduces their work to naught within no time. Such artists give material a different form for a short time, after which the material returns to its original state. Nothing more and nothing less. Art.

The most authentic work of art is us – man. Many people are a delight to look at; the whole world is their exhibition space. Moreover, every person is his own curator. It can't be a good thing: a work of art and its curator united in one. The problem is that we're most preoccupied with what other people think of us. Your cat couldn't care less if you're ugly: the only thing that counts is that your sweat smells good. A cat judges with his nose. But people judge with their eyes. Nearly all their attention goes to the outer layer, the skin, the body's shop window. What's so bad about that? Like the façade of a house, the skin, of all the organs, is the most exposed to wind and rain. This is certainly true where man is concerned, since he has no fur to soften the effects. It's no mean feat to protect one-and-a-half square metres of nakedness from the elements. Decay is inevitable.

Most people have their own studio for restoring their skin. What photographers do with a photograph, most women do before the photograph gets taken: touch themselves up. Before they dare show themselves in public, they've lathered on the cream and painted themselves with products that promise a more beautiful, youthful skin. They should be happy they don't have to treat every inch of their body! It's a simple fact that we hide most of our bare body under clothing. Young people cut out the occasional bits to show us what they have on offer, while old people are generally considerate enough to spare us the privilege. Such habits determine the appearance of high streets and shopping malls, which are mostly occupied by clothing stores and cosmetic firms. Nowadays, if as a middle-aged biologist you want

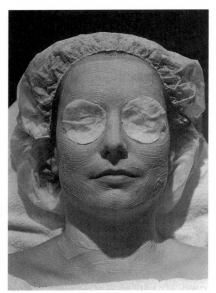

Cosmetics never reach the deepest layers

to buy something in a department store, you first have to make your way through two highly scented floors before finding something to your liking.

Many cosmetics are supposed to keep your skin young. In fact, you might just as well embalm your body. The irony is that the very skin you strive to keep young with creams is already dead. The outer layer is as dead as a dodo. There's nothing left to rejuvenate. Cosmetics never reach the deepest layers, whatever they claim. This is a good thing, because if any of those creams that you lay your hands on could permeate your skin, bacteria and dangerous substances could too. Using lipstick or mascara only for decorative purposes can be beautiful, but it's still no more than touching up parchment. And, in the end, restoration can turn on itself. Many old ladies in America – where, from childhood on, one cream jar more or less makes no difference – develop distinctly cosmetic faces, which can only be kept presentable by applying more cosmetics, until, eventually, as a European, you see only "the American woman" and not the woman herself. Generally, it's easier to tell a group of Chinese people apart than it is a circle of little old ladies from Florida. But such dangers pale in the face of the dangers of cosmetics from the past. In the days when fashion dictated pale faces, many women used to lather their skin with lead-white, adding antimony-black to their eyebrows for contrast. Lips and cheeks were then reddened with finely ground lice. This dye, cochineal, can still be found in fruit yoghurt, but today you're hardly even allowed to put lead white on your door frame, let alone your wife. And rightly so. Lead-white and mercury sulphate – for the lips – were fatal for the stunningly beautiful countess Maria Gunning, who died at the age of 27.

The only thing threatened by modern cosmetics is your wallet. The biggest problem facing cosmetics today is that no one is fooled by them anymore. People are so experienced in studying other people's faces – partly because of television – that they can see through the make-up as readily as a second-hand car dealer can see through the new paint job on an old Morris Minor. But the Morris is as unaware of this as the dealer's wife. The magic of cosmetics lies in the self-confidence one derives from them. It's always great fun to watch cheeky young 15-year-old girls, who don't know how

Decay is inevitable

to use colour very well yet, go out into the world with faces bursting with self-confidence.

Sooner or later, many people don't dare look at their mirror image until there's been a more substantial restoration. Nowhere has this desire been expressed more charmingly than in the Muiderslot, a medieval castle in Muiden near Amsterdam, where the little painting called *De bakker van Eeklo* (*The Baker of Eeklo*) delights the many busloads of schoolchildren who come to see it. In Eeklo, you used to be able to get your old head rebaked; while the decapitated head was being rejuvenated in the oven, green cabbage sealed the open neck wound. What came out was usually a great improvement, but occasionally the oven was too hot or not hot enough. In such cases, according to the guides at the Muiderslot, you came out either a hothead or half-baked. Naturally the guides have thought this up themselves, but that's the way it's supposed to be with a living legend. Originally, it wasn't bakers but blacksmiths who reconstructed women's heads. The heads were brought from far

The Baker of Eeklo. Copper engraving by Philibert Bouttats de Jongere (1656–1728)

and wide – on ships, wheelbarrows, men's shoulders and donkey carts driven by monkeys. Both bakers and blacksmiths were the butt of jokes about old women who had to be fixed up to "look like newly painted outhouses". But there was also a moral to the jokes: the heat of the ovens stood for purgatory, where man was purged before being allowed to start a new life.

Even though today's plastic surgeons have lowered their standards to a more earthly level they don't act simply in the name of vanity. The profession of plastic surgeon was originally invented to give war a more human face. What bombs and grenades tore asunder surgeons joined together again with needle and thread. Centuries later, these victims were joined by those victims from that war-with-no-peace that we call road traffic. Nowadays, a growing number of people consider their own lives a war, with their faces as the battlefield. The longer life lasts, the more effect gravity has on your skin. The only thing that helps defy gravity is standing on your head, or taking the lift. In the case of a facelift, the skin is pulled back until it's taut again, after which the extra bits are thrown away. The whole face can be treated like this, or parts of it, like the eyelids. It's usually well-known faces that dare to do this, although it can take quite a while before we know them again. Gravity has an even stronger hold on breasts. If they are to be given some form of expression again, either the contents have to be enlarged or the packaging reduced. Women attach great importance to the size of their breasts, men less so, as long as they're symmetrical. Men do worry, though, about the size of their penis, which, if need be, can be stretched by the surgeon. They're not so keen on facelifts. And by the time they could use one, they usually don't have enough hair left to use for covering the scars. They fiddle around a bit with hair implants, but these don't offer much consolation. Generally, there's little balding men can do – other than what Caesar did on occasion, which was to wear a wreath of laurels.

The demand for cosmetic surgery has grown rapidly, not only in America, but also in the Netherlands, where for many years it was reimbursed by the Dutch equaivalent of the National Health Service. To be eligible, the physical defect had to "fall outside normal variations in appearance". Committees met as if their lives depended on it. How much were your ears allowed to stick out, how big did your nose have to be to be considered abnormal? To have a facelift financed by the national health service, one had to look at least ten years older than one was; breasts could only be lifted at the expense of the State if "the nipples were at the same height as the elbows". There was no shortage of work!

Meanwhile, from behind the windows of many a restored Amsterdam canal house, not-so-very-old restored faces stare out at you. It's a symptom of the belief in – or rather the belief in the right to – a "makable" society. Your house no longer what it once was? Restore it! You're a year older? Off to the plastic surgeon! Nature has grown old? Create new nature! Whereas in 1973, 25 per cent of American women were dissatisfied with the size and shape of their breasts, by 1986 this figure had risen to 33 per cent, while during the same period their breasts hadn't changed at all. But not all countries are the same: Dutch women literally want a couple of grams less; their breast implants are a decilitre smaller than those of their American sisters.

It remains to be seen if the world is so makable. Old neighbourhoods that have been restored are usually more restored than old. Old people who have been restored are usually more old than restored. Fiddling with time leads to messes. Folds of skin begin to droop again, stitches become infected, connective tissue tightens around implants until they're reduced to doorknobs. As in art, restorers of skin and bodies are increasingly busy correcting each other's patchwork; one-third of all operations carried out by plastic surgeons are spent doing this. *Il faut souffrir pour être belle* – but far from everyone who suffers becomes beautiful. Plastic surgery is medicine in reverse: you go in healthy and

Above and right: Wrinkles as a trademark

you come out a patient. Nasty rumours say that anyone who wants to be cut up for the sake of appearances would be better off going to a psychiatrist; even nastier rumours say this puts a bit too much faith in psychiatrists.

All surgical interventions have their dangers. Movie stars put up with them as occupational hazards; they were the first to have their ribcages reduced, lips inflated and thighs liposuctioned. Marlene Dietrich's mysterious allure was partly due to her having had her back teeth removed early on in her career; this is what gave her those beautiful hollow cheeks. But there are some who don't worry about such things. While Jane Fonda has had herself almost totally rebuilt, Brigitte Bardot lets it all hang out. Simone Signoret even saw her wrinkles and bags as a trademark. And where would Michael Parkinson and Inspector Morse be today without their lines? Many of those who have had their eyelids lifted have turned out not to have so much depth after all. Today, certain breeds of dog, like the Chow-Chow and the Shar-Pei, are worth a lot of money precisely because their wrinkles are as melancholic as those of Alfred Hitchcock and Margaret Rutherford.

Many comedians think they're funnier if they're fat. Fatness suggests fun. A bumblebee is more fun than a bee; Pooh's posture alone makes him a like-

able chap. But, as happy as people are to see others fat, they're miserable if they're fat themselves. Beer bellies and fat thighs are associated with ageing and decline. So there's nothing left to do but go on a diet. In contrast to the repeated disappointment when the five kilograms come back again, there's always that repeated joy when those same kilograms are lost again. But no one wonders what happened to the kilograms. Five kilograms of person: that's a whole arm, half an amputated leg, one head less. More than 5 per cent of you is lost, gone, dead. He who diets looks decay right in the face. But where did you go? You burned yourself. When you diet, you cremate yourself, bit by bit, long before you die. Those five kilograms literally went up in smoke. Somewhere up there, part of you is floating around as carbon dioxide and water vapour. If every Englishman were to lose a kilogram after his holidays, the atmosphere would be enriched with 50 million kilograms of waste gas, and more than half-a-million gaseous compatriots would be contributing to the greenhouse effect. Whether you burn yourself or the plastic surgeon liposucks the blubber out of you and has it burned in the incinerator, the final result is the same.

Although you might be content to look at yourself in the mirror after being on a strict diet to see what isn't there anymore, you would be uncomfortable if, after an operation, you noticed that a whole leg was missing. It's much more difficult to say goodbye to a whole body part. Often it can still be felt. In the same way as a dog without a tail thinks he has to keep on wagging it, many people continue to feel pain in a leg they haven't had for many years. There's a simple explanation for this: there's no such thing as leg aches, backaches or stomach aches; there are only headaches. All signals from nerves come together – and turn into pain – in your head. The nerves in the stump of your leg might continue tingling, but the cells in your brain, which were sitting around doing nothing for want of a leg, automatically make new connections, so that if someone strokes your cheek, you might feel as if your amputated leg is being touched.

You can't walk on just one leg; you need two. Or four or six or eight, if you're a dog, an insect or a spider. The more legs you have, the steadier you are on your feet. Yet three or five or seven legs are as uncommon as one leg. A healthy animal has an even number of legs. There's a good reason for this: symmetry. You can chop me – and I can chop you – into two pieces with an axe, in such a way that one half is the mirror image of the other. The same can be done to carp, squirrels and cats. Thanks to bilateral symmetry, most vertebrates have two of most organs. These are the paired organs. Unpaired organs are generally located in the plane of symmetry: our noses, for example, or the dorsal fins of carp. Strangely enough, it's with an unpaired organ that pairing takes place. This is because we're not sharks; sharks have two penises. It's handy to have two of every organ. If one breaks down, there's always the other. If something's wrong with your left arm, you can still always use your right one. You can live quite well with one lung. And if you want, you can always give one of your kidneys to someone who doesn't have any at all. Seen in this light, you're only half yourself. The other half is a set of spare parts that you drag around with you all your life. The question is, which half is the real half? And why isn't it a complete set? There's no reserve heart – which is why it's most likely that this is what you'll die of.

But not all missing parts are equal. A person may have one leg, but never gets used to it. At a sawmill, on the other hand, no one would even notice if you had one finger too few or too many. That's why we have ten fingers instead of two. There's even one organ that can be amputated by the thousands – for no medical reason – on a commercial basis: hair. We complain bitterly when we no longer have it, but as long as it's still in good supply, we take it to the oldest surgeon around: the barber. He can't do much harm anymore because the only organ he can still put his knife into has already been dead for many years. So he can do whatever he likes. You've probably also undergone amputations of organs that *were* still alive. Almost every adult has had a tooth removed at some time during his life. There's a good reason for this: nature gave man too many dental elements. Our jaws are too small to house all our teeth; we're over-endowed. The first ones to go are the wisdom teeth. There's really no reason in the world for a jaw to be filled to the brim with teeth. Most herbivores have an empty space – the diastema. It would be pointless to have teeth there because teeth belong at the front, like the mouth of a pair of tweezers. Molars have to be at the back to do their work, because jaws, like nutcrackers, exert the greatest force near the hinge. Birds have no teeth at all. They masticate their food somewhere else – in their stomach, where it gets pulverized by hard ridges and little molar stones. But we can do that too, pulverize things outside of our mouths, albeit before, not

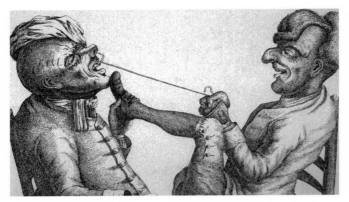

Nature gave man
too many dental
elements

after, swallowing: in the kitchen. There, amputated beef buttocks are chopped into little pieces and potatoes and artichokes are thrown into boiling water alive. By the time our food is ready, it's usually so well pre-digested that many people with false teeth often remove them so that they can really enjoy it.

Clattering dentures have given prostheses a bad name. They make your jaws shrink, laughing becomes awkward, kissing isn't what it used to be. False teeth are one reason why old people laugh so little and play the saxophone so badly. Young people try to keep their teeth for as long as possible – to avoid seeming old before their time due to noisy teeth. This is also the reason why half the people who need hearing aids don't wear them. Toothlessness and poor hearing are associated with old age. And rightfully so. But the same can be said of poor sight. Most people need at least reading glasses by the time they're 50. Oddly enough, this doesn't bother them. Unlike hearing aids or false teeth, glasses have become fashion articles with which to enhance your appearance.

So it's entirely possible to see fake parts in a positive light. From time immemorial, there's even been something romantic about them. As boys, pirate stories made our mouths water. The character with the wooden leg and iron hook may have been an invalid, but he was still the captain. We played pirate with the legs of chairs, the hooks of coat racks and patches over our eyes. In my mind's eye, I was also one of those people in the iron lung which they showed in the news bulletins at the cinema. Such a hideous thing reminded me most of an oven in which patients were roasted, with only their heads protruding. In reality, they weren't roasted but insufflated. A pump ensured that their chest expanded regularly, so that the lungs could fill themselves with air, despite paralysed respiratory muscles. How I would have loved to lie in such a steam lung! How I would have loved – those were the days of Meccano and build-your-own-radio kits – to build one! Your very own lung! With your very own door! Unconsciously, I appreciated the advantages of an iron lung over those of a wooden leg or an iron hand: it wasn't on your body; your body was in it. It encapsulated man, which made it as attractive as a car.

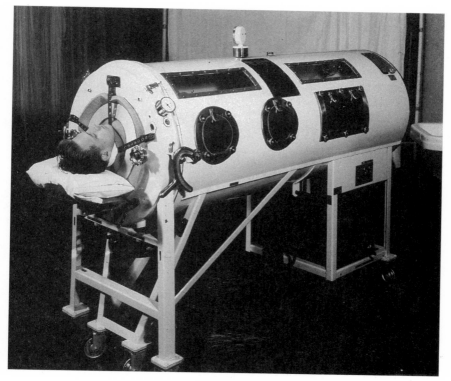

An iron lung

I used to peer enviously into shop windows full of magnifying glasses, hand-driven bicycles and chairs you could sit on so as to defecate in mid-room. If an invalid could manage, there should be no end to the fun a healthy young boy could have – with cutlery for the one-armed, little tubes in your trouser leg so that you never had to use the urinal again, and long-armed-pinchers to reach the top shelf! In the local cinema, we once saw a woman without a hand, who had had an adapter attached to the stump into which she could click an eggbeater, a potato peeler, a toothbrush or any other accessory she felt the need of. Such a living Kitchen-Maid showed there was still a great deal to be improved upon in man. Glasses help you to see almost as well as someone with good eyes; both eyes can see further with binoculars. An artificial organ can be better than the real thing. One day, sportsmen with the newest artificial organs in their bodies will burst out of the straitjacket of human anatomy. Why should an athlete be able to wear Nike shoes but not have Nike joints installed in his body? The initial resistance of sports federations will melt like snow in summer as soon as invalid sports-men start performing better with their new prostheses than sportsmen of only flesh and blood, and as soon as the public only wants to attend the Paralympics. By that time, normal people will also be able to choose between plodding along in ill health, with organs like the ones sold by the butcher, or having the very latest gadgets installed.

So, if the doctor is already inside you, why not let him make some improve-ments right away? Most people aren't interested in this. Getting better doesn't mean getting better than you were; it just means becoming normal again. Not God but other people are our heroes, with all the shortcomings that implies. Even if factories could make better organs than man, a person still wants a real heart instead of an artificial one, a real kidney instead of a fake one. We want original parts. This is why there's such a demand for donor hearts and kidneys. The demand is already so much greater than the supply that you'd have to set the traffic lights on green for weeks on end to accumulate enough dead people. In the same way that book restorers prefer old, weak linen instead of new, strong, synthetic materials, doctors prefer second-hand organs to fresh factory-made ones. Bad ones rather than imitations. That's restoration to a T.

There are no surgeons in nature. Yet many animals walk around with completely restored organs. The classic example is the lizard. As soon as its tail is in danger of becoming no more than a handle for the enemy, the lizard pulls special emergency muscles and the tail breaks off along a pre-formed line. While the enemy stands near the writhing appendage wondering what the hell is going on, the detached cab discreetly puts itself out of harm's way. After a while the tail grows back again. In some species, there are more specimens with regenerated tails than with original tails – as if regeneration were fun. So why don't carpenters grow new fingers? Because they're too highly developed. The more primitive an organism, the better it can regenerate: starfish continually grow new arms; crabs sacrifice their pincers to the enemy; bees sacrifice their sting. In all cases, the loose body part can continue to bother the enemy for quite a while. The difference is that the bee doesn't grow a new sting. Why should it? There are plenty of other bees, and thus other stings, in the hive. Hydras, or freshwater polyps, can be cut into ten pieces. That gives you ten hydras. Rather like striking cuttings. Asexual reproduction can be seen as an extreme form of regeneration. We humans don't have such magic tricks in our repertoire. Yet every moment of the day we regenerate vital parts of our organs. At the cell level, broken parts are replaced almost immediately. Although this kind of maintenance is less spectacular than a whole body part spontaneously regenerating itself, it is no less important. Wounds close automatically, worn-down calluses reappear, hair grows back, blood is constantly replenished. Organs and tissues aren't as permanent as they might seem. The skin seen in the mirror today wasn't there last month. Many cells – of the skin, the intestinal lining and the tissues that make red and white blood cells – divide every day, whatever your age. Your body is worked on until your very last breath.

There's a good reason why lizards grow new tails so quickly. It's because they have no tails to get in their way anymore. The same thing happens in trees: not a single leaf will grow out of a trunk for a hundred years, but if the tree is topped, buds will start sprouting out of the stump, freed of all the saps and structures that kept them dormant for all those years. In the stump of a lizard's tail – as in an embryo – shapeless cells start all over again with reconstruction. In a body which still has its tail, this can't happen. And in old city centres, old buildings have to be pulled down before new ones can be put up. Decay is essential for renewal. After decay, rebuilding is needed, but before you can rebuild, you need decay. Away with you! To meet this requirement, many cells commit suicide. Long before they break down and wither away as old cells, something changes in their nucleus, whereby an orderly self-destruction begins, often in groups, as in religious sects. The process starts in the embryo. Embryos are full of temporary organs, which are either needed to bridge the gap between two different stages or unnecessary as remnants of an evolutionary pre-history. Such organs must disappear, otherwise we would still be walking around with the gills of our fishy forefathers and hands that only mittens could warm. The only reason our fingers are separate is that the cells that joined them died off in time. In an adult body, cells commit suicide because they no longer function properly or because they were created for situations that never arose. In this way, antibodies against intruders must be rendered harmless at the right moment, so that they don't turn upon the body itself. If they refuse to sacrifice themselves in the name of the body's well-being, allergies or chronic inflammations – like rheumatic arthritis – can occur.

Despite all this breaking down and building up, an organism still manages to retain its shape and character. In this sense, too, cities behave like organisms. Canals are filled in but their beds remain; fortified cities retain their fortified structures; although surrounded by reclaimed land, you can still see the island in Ely. When Baron Haussmann had boulevards built through the heart of Paris in the mid-nineteenth century, the Seine kept flowing where it had always flowed since giving birth to the city, and Parnasse stayed a Mount. Character lasts. This is why it's so easy to love a city. Someone from St Petersburg loves his city more than his country. Precisely because he knows the old so well, he can enjoy the clash of the old and the new. Every construction

site is both a shock to the system and a sign that there's still life in the city –
albeit sometimes too much life.

Old houses are also at their best when they're the worse for wear: a crooked
staircase, one floor split into two, another adjoined to a third, a maze of
pipes from every possible period. But old houses are scarcer than old cities.
While urban planners have learned to live with organic growth, architects
plan the life story of a house the moment they design it, the way patresfamilias
once planned the lives of their sons and daughters from the moment they
were born. Generally, architects deliver what, in their eyes, is a ready-made
building, and then they take off, as quickly as they can, like sexually aroused
tomcats after mating. It's sheer Bauhaus. Indeed, form flows out of function,
but what happens when the function changes? All the novelties will be hope-
lessly outdated 25 years later, so that a new building will have to be built to
satisfy the new function – and once again crammed full of the latest novelties,
which in turn, 25 years from then, will be the new, old ones.

The alternative for demolition is restoration. Even in the land of everlasting
change – America – more money is spent on restoring and renovating old
buildings than on constructing new ones. Whereas in renovating, a building
is so heavily adapted to the new era that there's nothing old about it anymore, the idea in restoration is to retain the original character. The federal
government in America has contributed heavily to the restoration of many
such buildings. But, naturally, there are also many foundations, leagues,
associations and coordinating bodies to protect the cultural heritage and lure
the tourists.

Many restorers try to return houses to their original state. Damaged oak
beams are replaced with new undamaged beams, sculptures are cleaned,
crooked stairs are straightened (or not, as the case may be), colours are mixed
to reflect old renderings. After a long time and a great deal more money, the
building is handed over to its new owner in what is called "all its former
glory". A freshly painted yearstone shines down proudly from the façade.
But the date is a lie. There's nothing old about the house at all anymore.
Partly due to the restoration, all that was old has disappeared. Even if the

The ruins of the Huis ter Haar, in the Netherlands, drawn by Ad. Mulder, c. 1880

façade is meant to lean forwards and the window frames are supposed to be different sizes, there's no dirt embedded in the wood anymore, the old newspapers behind the wallpaper have vanished and there are no holes left for mice to discover. Children see this immediately; it's no longer a nice house. Only the form has been preserved. The old house has lost its soul; it has as little to do with itself as a mummy with the Pharaoh it once was. It's not even a faithful replica of bygone days. If canal houses were faithfully restored, they would still have wooden bogs and be without electricity. There are only minimal differences between the restoration of art and the restoration of monuments, so if their principles are basically the same, why are we allowed to put new wings on old windmills but not new arms on the Venus of Milo? Naturally, I'm happy to see old windmills turning in what remains of the Dutch countryside. But what about those wingless old mills with home-made shutters on their windows, the smoke from the wood furnaces trailing up out of their crooked chimneys and the hopeless disarray in their yards? Are there enough of these around? Is there an association to represent their interests? Such a "millhouse" is a thorn in the side of the inveterate monument-savers. Yet the wings of many of those beautifully restored mills wouldn't be turning today if their wingless foundations hadn't been occupied for many years. Re-use prolongs life. If a church can survive only as a supermarket, it will have to become a supermarket – and a power station will have to become a restaurant. It's even the fashion. Unfortunately, the restaurant purpose of the building, like the power station purpose before it, is often built so permanently into the structure that the restoration can no longer be undone – as is required in art.

So what's the alternative? If all you do is stand by and watch, you run the risk the building will collapse in a heap. Then and only then is there cause for action. Restoration is an emergency measure. If an old building has been well looked after, it no more needs to be restored than a normal person needs a wooden leg in old age. Well cared-for buildings can live almost forever, as long as there are people who want this to happen. And such people can only be found if they're allowed to adapt the buildings to suit their needs. Buildings must be adaptable; they only stand to benefit from it. There are good examples of this among monuments that haven't yet been restored. If they still have

Thirteen layers of wallpaper from between c. 1820 and 1910, applied on top of each other, found in an American house

twelve layers of wallpaper on their walls, you can just add a thirteenth layer. That way you'd be adding history. In a restoration, that very history gets removed. Someone who's well aware of this is the Dutch writer Geert Mak, who sang the praises of the Noorderkerk in Amsterdam in the early 1990s:

In a certain way the Noorderkerk is unique: it is one of the few monuments in the city that have not been restored. Whoever wants to know what a church looked like in past centuries has nothing to learn from visiting the shiny polished Westerkerk and the restored-to-death and given-new-life-as-a-party-centre Nieuwekerk. Only in the Noorderkerk do the doors still creak and is the bellows chamber still located behind the organ, as it was in the days when the organ pedals had to be trod by three burly men. Even the hooks on the doors of the toilets still date from the nineteenth century.

Let buildings live. Let landscapes live, together with statues, forests, whales, Eskimos and little girls. Live and let live. Take good care of them – not to conserve them, but to allow them to grow, along with their time. Replace roof tiles that have blown off, wipe runny noses, protect monuments from the elements, consult your doctor when necessary, clean what's dirty – and

The Noorderkerk in Amsterdam, hit by lightning at midnight on 1 June, 1979

cherish all these things. Make doors larger if you're rich and smaller if you're poor, nurture your house as it nurtures you, and stand by it during its final days. And when those days finally arrive, allow your house to fall apart at leisure, wall by wall. Watch how vines start to overrun it, let children play in it, grant wallflowers their pleasure, take beautiful pictures – and then build a new house, not too far away, so that the whole process can start all over again. Either a totally different house or a replica. Replicas don't have a very good reputation, yet the reason French villages are so beautiful is because their new buildings scarcely differ from their old ones. There are differences, of course, but they've retained their character. They're embedded in the same culture. Replicas? Yes. But all life is based on replicas. That your place on earth will later be taken by someone who was created in your image, a new member of your species – that's the oldest principle of life on earth. It's called reproduction. And it can be fun too.

5

OLD SEED

Nature is vast. To store it in a museum, you would need an enormous building. Great rulers love to be faced with such challenges. In London, the centre of an Empire where the sun never set, Queen Victoria had a temple built in honour of nature. The Natural History Museum most resembles a Romanesque cathedral. Awed by its high arches, many visitors used to remove their hats automatically when they entered it. In this gallery, built in honour of life, creation was laid out at man's feet, to the greater glory of Him who had created it. And of Her who had allowed it to be gleaned through theft.

Vastness is vulnerable and invulnerable at the same time. During the Battle of Britain, it would have been virtually impossible not to hit this megalomaniac museum. Early on the morning of 9 September, 1940, two incendiary bombs and one large oil bomb, discharged as ballast by the retreating *Luftwaffe*, crashed through the roof of the botany department. In no time, the Romanesque hall had turned into Romanesque ruins. The roof was destroyed by the fire. Fortunately, the dried plants had been stored in old-fashioned wooden cupboards, and wood is a poor conductor of heat. Behind modern steel doors, the herbarium would have been scorched, as punishment for the arrogance with which man had tried to save a piece of creation from God's eternal cycle of rise and fall. Apparently, the time wasn't ripe for that yet. The collection suffered most from water damage: paper turned to porridge, inscriptions faded, many original samples that renowned botanists had used to describe newly identified species were lost forever; numerous plants died a second death.

But where there's death, there's life. After the dousing by the fire brigade, several seeds came to life that had been brought from China in 1793 by Sir George Staunton. Their 150-year sleep in a dusty cupboard hadn't made them forget their mission. Unperturbed, they began to grow into silk trees (*Albizia julibrissin*). Their life was fulfilled after all. British botanists were amazed, even though the resurrection of seeds wasn't new to them. In 1856, they had sent 600 seeds from Kew Gardens to Australia, to help that country start its own botanical gardens. By chance, the seeds were planted only in 1906. Many of them sprouted, even though some were more than 100 years old.

The British Museum was hit by a bomb in 1940

It sounds strange: new life from old seed. Life's source is supposed to be young, fresh, unspoilt, without a history. But we're often attracted to the unnatural. It's a sign of wisdom that seeds don't squander their energy all at once, instead calmly waiting until the time is right. Seeds aren't stupid. Even without the clumsiness of English botanists and German pilots, they know how to bide their time. Those shed in the autumn wisely wait until spring before they germinate; in drier regions they wait until the rainy season. But seeds are like people: there are those among them that can't wait. Even in farming, some seeds germinate while still in the ear. Farmers dislike impatient seeds. They prefer those with a guaranteed dormant period which they can bring to life according to a fixed procedure involving a sudden change of air, light or water.

The canna lily (*Canna compacta*) seed is the most patient seed of all. During pre-Inca times in South America, Indians used to make rattle necklaces with them. These consisted of walnuts in which lily seeds had been inserted when the walnuts were still young – before they had grown hermetically shut. Seeds from a necklace dug up near Santa Rosa de Tastil in Argentina turned out to have been so well preserved that a plant was grown out of them in 1968. For people with a philosophical bent, puzzling over the age of these plants could be a pleasant form of mental gymnastics: are they 30 years old or 600?

The existence of even older living seeds was reported in China in 1995. Apparently the seeds had been germinated after 1,200 years. But, don't forget it's a holy plant we're talking about here: the Indian lotus (*Nelumbo nucifera*), which originated from the bottom of a dried-up lake where Buddhists had once cultivated lilies. In fact, their long viability probably had more to do with their thick walls and high level of L-isoaspartylmethyltransferase, an enzyme that repairs proteins, than with their holiness. To break the record of the Indian lotus, you'd probably have to go to the eternally frozen grounds of the Arctic. Seeds have little difficulty withstanding the cold. Most species survive easily in the freezer, because their food supply protects the germ against forming ice crystals. Tired of having to cultivate and water their plants year in and year out, many botanical gardens now have seed banks, where plants continue to live for many years as frozen seeds.

Yet not even seeds have everlasting life. For man, life is an ongoing struggle against death; the only real purpose of all that blood-pumping, digesting and thinking is to keep death at bay for a while longer. Seeds are no different. Although you can't tell whether they're dead or alive from the outside, using very sensitive measuring equipment it's always possible to see whether chemical processes are taking place inside that are indicative of life. In the absence of green leaves, which generate food from light and air, seeds must rely on their reserves. To do this, they need oxygen. Seeds breathe, just as we do. And if they don't germinate, they eventually breathe their last, just as we do.

So what's the story about those mummy seeds? As a young boy, I blushed when I read that the seeds of a Pharaoh – hundreds of centuries old – were still alive. I imagined myself sitting next to the Pharaoh's son at school. Later, it turned out they weren't talking about the Pharaoh's seed, but about the seeds of grain he had been given to take with him on the long journey after death. But even then, I found it incredible that wheat from Tutankhamun's tomb could be germinated 3,350 years later, and produce enough grain to fill a bakery with bread. Unfortunately, it, too, turned out to be a figment of the modern imagination. Grains of corn were found that were even 4,000 to 6,000 years old, and it was also possible to bake bread from them, but it wasn't very good bread. The grain was as dead as a doornail, almost completely carbonized. Because of their soft pods, grains can only germinate for a short time.

This story about the life-hungry mummy-grain came from Count Von Sternberg. Apparently, he sowed two real grains of corn from a real sarcophagus that really came from Egypt, and then watched with amazement as they really sprouted. The only thing that wasn't real about the grains was that they were really old. Instead of being harvested 3,000 years earlier, they had been harvested – at most – three months before. The sarcophagus from which they originated had been used as a feeding trough for the horses of the King's viceroy. It only took a money-hungry guide and a gullible count for a new myth to be born.

But my youthful misunderstanding wasn't so mad after all. From rye seeds comes rye and from common comfrey seeds comes common comfrey, so that you might say that Pharaohs come from Pharaoh seeds. Seeds are seeds, after all. The only thing that differs is the soil. Rye and common comfrey can survive in cold soil, human seeds need the mother's warm womb. The only seed a gardener doesn't sow in his garden is his own. No baby biologists will grow out of my seeds if they're planted in fertile soil; they'll do better in a fertile woman.

That's what was thought until recently. Man supplied the child; woman was simply the living cradle. Until only a few centuries ago, the discoverer of the human sperm cell, Antonie van Leeuwenhoek, was convinced that it contained a complete human being. Another Dutchman, Nicolaas Hartsoecker, even observed it under the microscope, knees bent and arms folded to save space in the tiny capsule. Propelled by its flagellating tail, the little being hastened into its mother.

It wasn't until the mid-nineteenth century that it became clear that the father and mother each provided half of the raw material for a child. Only then was it realized that plant seeds are very different from human seeds. The seed of a plant houses a tiny being that lives off the food supply it was born with until it can stand on its own roots; it already has a father and a mother. The only animal-like thing that a plant seed can be compared with is a fertilized egg. Such an egg also houses a little being: a child from two parents, with a reserve food supply, and encased in a hard shell, which later, when the little being tries to find its way out, it will regret having.

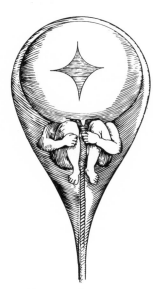

Nicolaas Hartsoecker saw a complete little person in the sperm cell he observed through his microscope in 1694

In this comparison, human seed – or semen – is equivalent to the powder on the plant's stamen. It only has a father and still has to unite with an egg cell before it can develop into a little being. Since plant seed was known for centuries before human or animal seed, it would be more appropriate if man were to call his sperm "pollen", but this evokes visions of unsolicited services by worker bees, with their horrid probosces and stings. What's more, a man would have to call his rod a "stamen". As long as he's not prepared to do that, the result will be a confusion of tongues more deserving of feminist indignation than if an office full of secretaries was addressed as "Dear Sirs". But we're used to a confusion of tongues in sexual matters. My guess is that you, too, have always thought that the pistil of a flower was the male organ.

A human being can't grow out of a human seed without the help of an egg cell, but a tree seed can produce a whole tree. Every cell of a tree's seed contains a complete blueprint for the entire tree. The information is stored in the spiral staircases of the DNA. It's strange: here we are in the era of databases and missile technology, and we talk about the secrets of life in terms of old-fashioned metaphors such as spiral staircases and blueprints. Spiral staircases date from the days of medieval castles and the blueprint was discovered by Sir John Herschel in 1842. Who still knows that blueprints were yellow, not blue, during the first hours or days after they were made? Certainly not young geneticists. But this hardly prevents them from talking about blueprints all the time.

If a seed is the blueprint of a plant, a blueprint is the seed of a building: all the data are encoded in it, with slightly different data for each building. The difference is that the seed of a building is not destroyed as a result of its own realization. Vaults in city halls and architectural offices are full of blueprints for buildings that have been around for a long time or that have even been torn down, as well as for buildings that never saw the light of day. Tens of thousands, perhaps even hundreds of thousands, of blueprints lie in drawers around the world awaiting fulfilment, like sperm in a sperm bank. The chances that this will ever happen are nil. Architects always think newer is better; they're all too eager to allow other men's seeds to shrivel up and die.

So we needn't be surprised to learn that there are still about 32 Natural History Museums in London: that is, in the form of blueprints. Each one is detailed enough in its own right that, if a future bomb were to hit its target, the present museum could be replaced with a design from the same period. This abundance is a result of the competition that was held for the museum's design. Architectural competitions were the order of the day in England at the time. The Bank of England, the National Gallery and the Houses of Parliament also came into existence as a result of rivalry. Well-known prizewinners else-where in the world are the White House, the Eiffel Tower and the Amsterdam City Hall. People only see the single oak tree; no one mourns the many acorns that were lost for the sake of that one tree.

The longer a blueprint lies in a drawer, the slimmer the chances the design will ever be realized. Garden-lovers are familiar with this phenomenon from real seeds. If 100 young seeds produce 99 seedlings, 100 middle-aged seeds will only produce 50 or 60. What's more, these seedlings will grow slowly and there will be more misformed specimens among them. Old seed is weak seed. But the same isn't true for human beings: many a grandfather has become a father. About one in every 5,000 postwar Germans was born to a father older than 60, and one in every 30,000 to a father older than 70. The oldest known young father to be found standing at the side of a cradle is 94.

Old men are dirty old men. When you see them sitting in front of the windows in homes for the elderly, you wouldn't think it but every day until their very last breath, they produce millions of seeds right under the eyes of the matron. Of course, their sperm is a bit weak – there are more and more rejects among them, the flagella have become stiffer – but there are still enough high-spirited spermatozoa to inseminate the matron. That is, if she's not too old for it.

If the principle of excess is proven anywhere, it's in man's sperm production. Sperm are as plentiful in a man's scrotum as seeds are meagre in the over-priced seed packets you buy at garden centres. Though only one spermatozoon can eventually cross the finishing line, each time anew millions of little spermatozoa stand poised at the starting line, flagella wagging. And this is nothing compared with the seeds of many marine animals. Males of most marine species don't have a spout as handy as ours for bringing the seeds as close as possible to their goal, so they simply dump them into the sea without much ado and leave them to find their own eggs to fertilize. To give the spermatozoa a realistic chance, they produce them in astronomical numbers. Since the creatures that do this also occur in astronomical numbers, the best way to describe the sea – where so many of us spend our expensive holidays – is as diluted sperm.

It's safer on land. If a land animal were to discharge its sperm onto the ground, the sperm would have dried out long before it was able to do any inseminating. Yet even on land, we still wade through clouds of sperm: it sticks to our hair and forces its way up our noses. Plants that are dependent on wind for their reproduction disperse their sperm so generously that many people get hay fever from it. Such pollen doesn't need flagella. Wind provides free transport – up to more than 400 kilometres a day and up to an altitude of 6,000 metres. The pollen of alders, birches, oaks and grasses has even been found in the mid-Atlantic. Since grains of pollen are sensitive to the ultraviolet rays of the sun, the pollen has lost some of its vitality, but, given its abundance, it can afford to suffer.

Old scrota think along similar lines: several hundred thousand fewer or more deformed, senile or otherwise defective sperm cells make little difference. Sperm cells are produced in such profusion they wouldn't even be missed as standbys. The purpose of this excessive excess is to beat other sperm to the draw. The more sperm a man pumps into a woman, the smaller the chances a rival's sperm will be able to claim victory. And the greater the chances of unfaithfulness, the greater the sperm production will have to be. Within a single group of animals, the degree of infidelity from species to species can be determined by the size of the scrota. Usually these are hidden within the body, out of sight, but in some animals of our kind – the apes and monkeys – they dangle about outside for all to see. It only takes a bit of measuring to reveal that the scrota of chimpanzees are four times larger than those of gorillas. So gorillas are the most faithful. While a male chimpanzee does it hundreds of times a year with dozens of females, gorilla males jealously guard their few partners, waiting for those rare moments – between

Many old goats are still searching for green leaves

pregnancy, giving birth and nursing – when the females are willing to bestow their favours. And what about people? How faithful are our scrota? In terms of size, ours are between those of the chimps and the gorillas, albeit leaning towards the gorilla side – the side of faithfulness. But there's enough room for infidelity until a ripe old age. This has – gratuitously – been proved by a study at the University of Münster. It turned out that a group of old men between the ages of 60 and 90, with their three millilitres of sperm, produced not one drop less sperm than a control group of virile young men. And, although their sperm cells were slightly less spry, they penetrated the egg cells allocated to them just as eagerly. Moreover, there were also more sperm cells in the old men's sperm than in the young men's – 125 million per millilitre compared with 75 million – but that's because old men keep their sperm for longer. But their sperm already proved itself long ago. Old men who are still bent on a bit of proliferation can safely leave the distribution of sperm to their sons.

Old men make young sperm. Fresh sperm is produced after every ejaculation, like bread at the bakery. Old men spout fresh sperm just as old bakers sell fresh bread. It's different where women are concerned. Women lay old eggs, even when they themselves are young. Every month they lay an old egg. It's so small you can't see it. But it doesn't have to be big because, unlike a chicken egg, it scarcely has any yolk to house. If the egg develops into a

Just as old bakers can sell fresh bread, old men can produce young seed

baby, there's enough food for it in the womb. A million human eggs – only two millimetres long each – fit into a single chicken egg. Yet, as unbothered as a woman is by the actual laying of the egg, she is bothered by the monthly bleeding that accompanies it.

Strangely enough, from the moment a human egg is laid, it is lost to fertilization. And, if an egg is fertilized, it has to stay inside in order to grow into a person. But the egg from which a baby emerges is also anything but fresh. A woman's eggs are about as old as the woman herself. They were made when she was still in her mother's womb; she cannot make any more after that. Throughout her whole life, a woman has to make do with the supply of egg cells that she inherited from the embryonic phase. This is the reason she doesn't need a lot of them: to be able to produce one egg a month during her fertile years, a woman only needs 500 eggs. This is nothing compared with the billions of sperm a man produces during his lifetime. Yet women, too, work with excess.

Women are born with no fewer than one million eggs in their ovaries. By the time they're 20, a quarter of these are left and by the time they're 40, only several tens of thousands remain. During this cell massacre, the bad ones are probably weeded out so that the good ones can survive. Otherwise, 20 or 30 years after they were formed, many more of them would be defective. The maximum shelf-life of an egg cell depends less on the degeneration of genetic material than on the manner of conservation. Egg cells are stored unripe. Their development ceases halfway through cell division – before the chromosomes of the mother cell have been divided between the two daughter cells. By the time 30 or 40 years have passed, the dividing mechanism has deteriorated so much that one daughter cell is sometimes given one chromosome too many. In that case, the child can develop Down's Syndrome. The chances of having a Down's Syndrome child therefore increase for women in their late forties – from the normal rate of one-in-700 to one-in-ten. There's nothing wrong with the mother herself, it's just that the date of her egg cells has expired. Fortunately, the body has a safety mechanism. By the time the chances of bearing a defective child become biologically unacceptable, the ovaries no longer produce enough hormones to ripen the egg cells and the woman stops laying eggs. She is now in the menopause.

Hens also stop laying eggs. Fewer of their eggs ripen each year; while any self-respecting hen lays 175 eggs in her first year, she lays only 125 in her second year and fewer than 100 in her third. Dinner's served! But most animals never reach the menopause. They die too young. Something resembling the menopause has only been observed in zoos and laboratories – in rhesus monkeys, pig-tailed macaques and chimpanzees. In the days when we humans still ran through the forests stark naked, we didn't reach the menopause either. All life after 40 is an "encore" given to us by culture that nature hadn't counted on. Biologically speaking, though, it's still a good thing nature didn't give us an extra fertility package for that repeat performance. Humans are not only unique in the menopause, but also in the long period during which they care for their young. It's a bit late to have your last child at the age of 50 if you want to be healthy while raising it. Accordingly, the latest known age at which a woman has borne a child through natural processes is 59.

Menopause is the opposite of puberty. Yet there's a striking similarity: women do strange things during this phase of their life. In both cases, hormones play a role. In the same way as the sudden flow of hormones gives rise to all those problems we call puberty, so the turning off of the tap has its effects. The most striking are hot flushes. From time to time, blood suddenly rushes to the head, which then feels as if it's on fire. This peat fire – which can last for up to five years – flares up repeatedly and lasts anything from a few seconds to several minutes. It's initiated by the brain, which doesn't know how to deal with the lower levels of oestrogen and so,

Women do strange things during the menopause

just for the fun of it, makes the blood vessels in the head and neck expand. But if the brain becomes confused from the lack of oestrogen, why doesn't this bother young girls, too, before they reach puberty? The answer, strangely enough, comes from adult men. Men with prostate cancer who are treated with oestrogen also get hot flushes when the therapy stops: the nerves that regulate the supply of blood to the skin have become addicted to the oestrogen and are kicking the habit. It is significant that the reduction in the level of oestrogen is also regulated by the nervous system, but what causes the nervous system – where the trigger for the menopause is – to do this is still unknown.

Other side effects of the menopause are much better understood. Aside from feeling flushed, many women in the menopause feel unhappy, tired and jittery. They start sweating, wake up much too early and start to think that the Creator might be sexist. One turns to drink, another – for want of better – turns to men, yet another can't be torn away from the mirror. Only the odd woman experiences the end of fertility as a form of liberation. And these are just the temporary symptoms. Some symptoms are more long-term: the deterioration of the reproductive mechanisms, the continuing threat of incontinence and the wrinkling of the skin. Hair that disappears from the three body crevices reappears as a moustache, and the calcium that seeps out of the bones seems to harden in the arteries. All of these symptoms can be eased to some extent by supplementing the lack of hormones with medication. But, increasingly it seems, a price has to be paid for the fact that the same hormone can have so many effects at the same time. Hormone replacement therapy, for example, can increase the risk of breast cancer and vascular disease.

Ever jealous of what women have that they do not, men now claim to have their own menopause: the penopause. In their forties, they also suddenly start not feeling very well and think they too should be allowed to act strangely. Given that fertility doesn't end suddenly in men, the penopause seems more like an excuse to have an affair than a phase of life. Instead of physical symptoms like hot flushes, feelings like "Is this all there is?" dominate. As punishment for their feelings of self-pity, many middle-aged men make themselves go jogging. They plod along through wood and vale, their fears in hot pursuit. If, at the same time, they were to pay more attention to the birds, they might even witness a genuine male change-of-life. After about four years, the sperm glands of quails wither away to nothing. Unlike human males, who use the penopause as an excuse to seduce their secretaries, quails become less sexually active as a result. In the absence of male hormones, the males lose all interest in the females.

Of course, hormones in human males also diminish, but they do so much more gradually. Most men were so saturated with them during childhood that they have enough left to last a lifetime. But the fact that they still decrease is reflected in waning powers. The older a man gets, the less he wants. And what he wants less he can also do less, according to the famous

Under 25 – Twice daily
25 to 35 – Tri-weekly
35 to 45 – Try weekly
45 to 55 – Try weakly
55 and on – Try, try, try.

An old prick leaks. The blood that's supposed to keep it stiff seeps away through the veins. And all this, while the arteries that are supposed to pump it full of blood have become narrower and more rigid. But the main culprit is the nervous system. The nerves in the penis react less to being touched and the brain orders too little testosterone to excite the penis fully. Women, by the way, are less bothered by this than men like to think. Thanks to their reduced libido, men have to take more time for the deed and that's a nice bonus for many women, who can now be satisfied at leisure.

Sex is like eating. Although you have less of it as you get older, it's still nice. Health permitting, old people continue to enjoy their portion of sex, even though it may not be as passionate as it once was. Until a ripe old age, the elderly ogle, play footsie and venture to make love. A study carried out by Ingmar Skoog in a home for the elderly in Göteborg revealed that 10 per cent of the residents above the age of 85 had had sex during the preceding year, while 30 per cent had had sexual feelings. Homes for the elderly don't like this. It creates greater demands for privacy and makes the old-timers edgy. Jealousy brews. The personnel often try to suppress it by avoiding talking about sex. The result is that the old people often don't know what's physically for the taking at their age, and for the second time in their life they're the victim of inadequate sex education. Sometimes they take their frustrations out on the personnel. That'll teach them!

As difficult as it is to measure the decline in physical lust in old people, it turned out to be easy to measure in old rhesus monkeys. A female who only had to lift a latch to join the male in the cage next to her, lifted the latch less and less frequently as she got older. The need wasn't so strong anymore. The study doesn't mention what the male thought of this. Probably his need diminished too, although perhaps not as quickly as hers.

One thing is certain though: human males don't have much of a sense of humour when it comes to witnessing the decline of their own bodies. It's as if the last thing they have to hold on to goes limp in their hands. They consider it the body's ultimate insult that the man in them dies before they themselves do. Something has to be done about it. Aphrodisiacs are shipped in. Sexual organs and sexual products are good candidates; sheep's balls and caviar are dished up, together with eggs – and an enormous wink. Lookalikes are equally acceptable. This explains the popularity of asparagus and bananas on the one hand, and mussels, prunes and figs on the other. Animals with penis-shaped organs are the biggest losers of all: the rhinoceros has been brought to the brink of extinction because of its horn. Poachers kill the rhinoceros so they can saw off the horns; conservationists saw off the horn so they can save the rhinoceros. Subsequently, the poachers also kill rhinoceroses without horns so they won't have to chase them in vain. Impotent Chinese men won't settle for cow's horn; it's not exotic enough. Eagle claws and tiger balls, on the other hand, do have that exotic aura. In Europe this was once even true for products like potatoes and cacao, which are very common now but were very exotic then. In the Marquis de Sade's stories, partners very often recover their stamina with cacao: "After his orgy, the King of Sadigne offered me half of his chocolate." Other products that literally made

Conservationists saw off the horn to save the rhinoceros

one hot – such as peppers, sambal and Spanish fly – were also very popular.

Spanish fly isn't an exotic fly at all but a beetle indigenous to much of Europe. To make it into a love potion, it's dried, crushed, then dissolved in a bit of alcohol. If you don't notice any effects after swallowing it, you've been lucky. Hopeless though it is in stirring up your lusts, even a small dose of it can wreak havoc on your kidneys. The active ingredient, cantharidin, is an aphrodisiac – but primarily for the female beetle, who is given a large dose of it during copulation. If Spanish fly ever made the imagination run wild, it's thanks to man's best aphrodisiac: his imagination. This is what releases the real love juices – the hormones. Until a ripe old age, almost every man is capable of thoughts so disgustingly obscene they could raise a builder's crane.

If dried up balls can age you, juicy balls can make you young again. These were roughly the thoughts of the eminent French physiologist Charles-Édouard Brown-Séquard. In 1889, at the age of 72, he announced that he had injected himself with an extract from the crushed testicles of a family pet; he claimed he felt better for it. In the 1920s, the American John R. "Doc" Brinkley and the Russian Serge Voronoff picked up this thread again. Brinkley became so wealthy from goat ball implants that he nominated himself candidate for the governorship of Kansas. Serge Voronoff earned a fortune by implanting slices of monkey testicles into little old men, who were as gullible as they were poor lovers. The men were happy, the monkeys less so. But Voronoff went too far. With the help of testicles from young horses, he tried to squeeze the last drops of sperm out of old but still famous studs. Insensitive as old studs are to the power of suggestion, Voronoff's game was over. Exit glandular therapy. The only question that remains is why old women have never allowed themselves to be injected with ovarian juices.

Although the longer you live the less sex you have, the less sex you have the longer you live. This line of reasoning washes better than Brown-Séquard's. Sex is expensive, risky and tiring. He who chases women or seduces men neglects his daily business. During copulation, you're vulnerable; after copulation, new duties call. Worries rear their ugly heads, mouths need to be fed, sacrifices made. And if you succeed – if you give reproduction your best shot – what do you get for it? You've given life to your own rivals! Mothers stay pregnant, fathers wear themselves out. Compared with this, eunuchs are the picture of health. Just watch the fat ox go about his business while the steer jealously guards his herd; tuck into that fat capon while the cat decimates the pitiful one-day cockerel; and ride at leisure on your gelding past the squeezed-dry stud on that stud farm. Even people have something to gain from castration. If you castrate a

The eunuch Carlo Broschi, nicknamed "Farinelli" (1705–1782)

young boy before puberty, he'll stay calmer, keep a clear complexion and be much less likely to commit a crime; he'll also never develop a beer belly. In short, he'll become more feminine. As well as disadvantages, castration has one great advantage: eunuchs live longer. Men have seldom been convinced though. But castration in a mild form has been used for rejuvenation. In Vienna, during the 1920s, you could have your seminal tubes tied at Professor Eugen Steinach's. This left you sterile but was supposed to stimulate the testicles to produce more hormones and thereby rejuvenate their owner. In theory, there's something to be said for this; in practice, it turned out to be hogwash. Today, the operation is performed on men who don't want to have any more children.

A less rigorous alternative is to leave your equipment intact without using it anymore. That's what priests vow to do; and monks do, indeed, live longer than their married brothers and cousins. The effect of total abstinence on the life span of women is more difficult to determine, but it goes without saying that giving birth reduces life expectancy. Abstaining from sex has other advantages too. Your career finally gets all the attention it deserves, as every anthill and beehive shows. Freed from sex – which dominates many of our activities and, at times, all of our thoughts – the (female) worker bees can work from dawn to dusk for the common good. What we call diligence in insects is nothing more than chastity. But, surprisingly, the worker bees don't live longer because of it. With a life expectancy of only a few weeks, they're no match for the queen, who can live to be several years old. The same special food that makes a queen out of a larva ensures that she will far outlive the other larvae. There's a reason why royal jelly is found in so many rejuvenation pills – although, they will no more help us to live longer than they will enable us to fly. The question is, then, whether the queen really does live longer because of it. It's equally possible that she reaches the normal age and that the worker bees die earlier because, before long, they produce much less for the hive than it costs the hive to produce a new worker. The worker bee must die on time because the new worker bees are much cheaper to support than the old ones. Male bees don't get old either. Most of the drones are simply killed off as soon as the queen bee is fertilized, and the drone who had the honour has to pay for it with his life: his penis breaks off in her body and what's left of him falls to the ground in a dying heap.

If you have to die anyway, dying from love would seem to be the best way to go. Many poets have sung the praises of *la mort douce* – death due to heart failure during lovemaking, when someone who shouldn't allow himself to be titillated has surrendered to the most titillating of titillations. In the wild, death from love is very common, not only among insects, but also among fish, like salmon. By the time a male salmon has left the sea and travelled far enough up the river to deposit his milt on her roe, it's so exhausted that it dies a short time later. That salmon are dying out is thus partly their own fault – who would be fool enough to make such a long journey for such short-lived pleasure? – but riverine pollution is giving them a helping hand. The male saiga antelopes that roam the steppes around the Volga and the Caspian Sea don't actually die from mating, but it wears them out so much that they become easy prey for wolves. Another mammal, the marsupial mouse *Antechinus stuarti*, also seems to age quickly after mating. Like a God-given warning to us sinners, the males, scarcely a year old and exhausted after an excessive group orgy that lasted several days, collapse and die – felled by all the excitement, suffocating from their own stress hormones.

The extent to which love and death belong together is symbolized in our culture by the arrow Amor uses to pierce lovesick hearts. Sooner or later you'll die of love. So it's not so illogical that a wedding car also often serves as a funeral car. The only real way to see whether it's driving in the name of love or death is from the flowers it's carrying. This can't be a coincidence either – that the love organs of plants are killed for our loves and our deaths. There's no clearer way for love to point to death than this.

None of this deters old goats from lusting after green leaves. Young goats searching for brown leaves, on the other hand, are a less common sight. The laws of reproduction dictate that young is supposed to fall for young. Gerontophilia, a love for those who are much older, is considered a deviation, but not a serious one. Usually it involves young girls falling for older men.

You seldom hear such men complaining about the fact that their twilight years have become gilt-edged. The women, too, can count themselves happy; they were probably looking for security and support, maybe even a father. They seldom have to feign headaches and what they miss in passion can always be compensated for in money. In love triangles – once-popular caricatures in which old men are copulating with young women – it's not a coincidence that a purse is always depicted. The third party in the triangle is usually a young man, who makes a fool of the old man. The woman agilely slips him the key of the chastity belt that the old fool makes her wear, and the young lover knows exactly how to slip it into the lock. The love triangle thus reveals not only her powers over her lover, but especially the powers of young over old. In the scheme of things, age differences – if not too great – are entirely normal. Just as a man is nearly always taller than his wife (even if she herself is tall) and seldom shorter (however short he may be), he is almost always older than she is. There's a lot to be said for this early on in marriage – an older man protects better, a young woman bears better. But, in the end, the woman becomes a widow earlier than necessary – because women outlive men anyway.

If you want to live to a fine old age, be a woman. This seems to be the best reason for having yourself rebuilt from a man into a woman. On average, women outlive live their husbands by five to ten years. Manliness is a fatal illness. This is partly attributable to an unhealthy lifestyle. The more cigarettes women smoke, the faster they drive their cars and the higher they climb the social ladder, the quicker the gap will close. But manliness is also a congenital illness. As long as men and women differ in the heights they can jump, the rates at which they go bald, the results they achieve at chess and the age when they begin to act like adolescents, there will be differences in the moment when they die. If the Olympic Games were to introduce a "survival" event, different rankings would have to be introduced for men and women, as they have been for the sprint and the shot-put. The trick question is of course: how important is the genetic component and can anything be done about it? And if women live longer than men, why are there always so many more women in the doctor's office? Are women more

Man's decline
illustrated in a
nineteenth-century
Friesian folded
letter

fragile? Or do they live longer because they go to the doctor more readily?

In all age groups, women have higher medical expenses than men. In fact there's nothing more costly for the healthcare system than old women. But it's money well spent, because women are easier to cure. Typical female disorders such as rheumatoid cellulitis and lupus seldom kill, however awful they may be. But it's a different story where cancer of the typically female organs, such as the womb and the breasts, is concerned. The average age of women would be even higher if they rid themselves of these organs at the right moment. According to British gynaecologist James Owen Drife, women would do well to have their breasts removed before cancer developed in them. Obviously, most women wouldn't even consider the idea. They would literally put their lives in the balance for their breasts. They need them. After all, men don't chop off their penises, do they? No. And rightly so: there's very little penis cancer. Cancer in men is usually in the prostate gland, the stomach or the lungs. But before they die of any cancer, most men have already succumbed to muscular dystrophy, haemophilia, heart failure, brucellosis, asthma or one

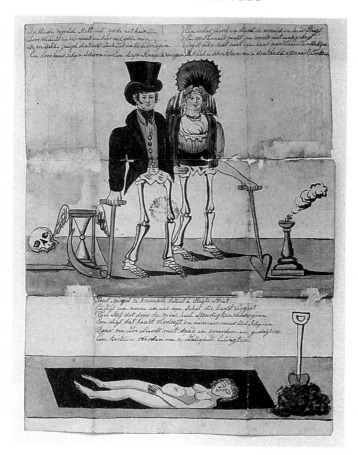

of the many other illnesses that afflict men more than women, or to accidents of course. Only the sensitivity about old age is the same in men and women. Old men don't die earlier because they're older; they die earlier because they're men. And they've been dying faster than the other sex all their lives, even before they were born.

Men are not only easier to break, they're easier to make – disposable articles. For every 100 girls, Dutch women bear 105 boys. This defies the laws of probability. Whether you're a boy or a girl is decided by fate. Your father's sperm plays the role of a die, but a die with only two sides: an X side and a Y side. Together with the X-chromosome of the egg cell, it forms either an XX (girl) or an XY (boy). Because there are as many sperm cells with an X-chromosome as with a Y-chromosome, 100 boys should be born for every 100 girls. In reality, though, many more Y-sperm cells win the race than X-sperm cells. One-and-a-half times more male fertilized egg cells are produced than female fertilized egg cells. But those who run fastest die first. From the very first cell divisions, men are less hardy than women. If they get through the embryonic phase, boys cause more problems at birth than girls, due to their heavier bodies. This disqualification race continues throughout life. By about the age of 30, so many men have died that they've completely lost their headstart and the ratio is about one to one. Unhappy with the balance that has been restored, they keep on dying steadily after that, until they're all gone, long before the women. By the time they're 70, there are twice as many women as men, so that homes for the elderly resemble homes for women. Ageing is effeminizing.

For a long time, the cause of this was sought in the sex chromosomes. It was thought that if something was the matter with a gene on the X-chromosome, this would be compensated for in women by the healthy opposite number on the other X-chromosome. But men don't have a second X-chromosome. And, indeed, muscular dystrophy is caused by just such a weak X-chromosome. The theory doesn't hold water, though, because in women only one of the two X-chromosomes is expressed in each cell. So, in this sense, women are no better off than men. Moreover, among birds and butterflies, for example, there *are* males with two X-chromosomes and they don't get any older than the one-X-chromosomed females. One good look at nature shows that a variety of factors determine who lives longer. If the males have to fight hard to win the females, they usually pay the price. If the egg-laying and parental care turn out to be exceptionally rigorous, the females are the first to die. Perhaps more important than death due to injury and miscarriage, however, is the overdose of hormones needed to allow for distinct sexual roles.

Among human beings, females outlive males. The question is whether the female humans should be happy about this. All too often they spend their last healthy years taking care of an ailing partner, after which they then spend their real last years alone moping – and robbing each other of the few remaining old men. While old men enjoy a certain degree of popularity, old women are held in low esteem. Only old men are considered prize attractions in the local pub. In literature, old women tend to be depicted as witches, angry mothers-in-law, meddling types behind the scenes, who only come out of their houses to drag their men into them. If only they drank more, they'd not only be more fun, but more importantly, they'd live and die at the same rate as men.

As long as women spend their whole lives trying to look young, they'll have to pay a heavy price when this role is lost to them. If they're lucky, they can still play the grandmother, but without grandchildren they can't even do that. In the end, the most popular and peaceful role left for them is that of the old woman in her garden – a role that suits old men too. A prudent garden centre uses an old gardener in its advertisements. Happily hoeing and weeding away in a straw hat – that's the way we like our old folk best. Harmlessness personified! But take a good look at what they're doing: growing seeds, propagating sexual organs, fertilizing and pollinating. What could be more wonderful if you're old than to give birth to new life! But just keep your fingers crossed that you'll live long enough to be around to watch it die.

6

THE RAVAGES OF TIME

The mouth is the death of the teeth. It's not a safe haven your palate offers them, but a torture chamber, a battlefield, a hellhole, a death trap. Your teeth are continuously besieged by the rabble that cohabit with plaque. A teeming metropolis, that's what your mouth is, full of slums and alleyways that are becoming more decrepit by the minute. Like befuddled pub-crawlers at the foot of old cathedrals, streptococci urinate against the crowns of your teeth, with a corrosive acid that no cement can withstand. Large cavities develop so quickly that before long the revellers fit into them perfectly. Every day chemical warfare with toothpaste claims countless victims among the slum-dwellers, after which blood and saliva vigorously attend to remineralization, but in

no time unbridled procreation restores the metropolis to third-world proportions. Tooth after tooth goes to seed, molar after molar succumbs to the jostling of microbial orgies, until only a single molar rises up out of the battlefield. Aristotle believed this was proof of how perfectly the world was constructed: at the very moment you died and no longer needed them any more, nature ensured your teeth were finished. Only help from outside can change this. Dentists can turn the tide,

The mouth is the death of the teeth

but that's not the same as preserving your teeth. Many people have more lead and mercury in their mouths than is legally permitted in the ground their houses are built on. The best thing for a tooth is if its mouth dies young. Free of sugar-loving streptococci, teeth last much longer in a corpse than in a live body. Long after the tongue has been eaten away and the uvula has rotted, your teeth still rattle around, alive and well, at the bottom of the coffin.

Yet, while teeth can manage fine without mouths, mouths are lost without teeth. Teeth helped vertebrates and anthropoids gain supremacy on earth: thanks to teeth, they can grind their food so fine that all the energy is released from it. Since reality is too large to consume in one bite, rats, sharks, locusts and vultures hack it into pieces first. Bacteria and fungi finish the job, until everything's gone. But when our teeth decay the incriminating finger invariably points to time. The ravages of time, that's what did it.

This is pure slander. The ravages of time don't even exist. Time doesn't ravage; time does nothing but pass. What gnaw are mice and worms and fungi and consciences – often at our behest. Such was the case several years ago at a police station on the

Little gnawers are suitable as paper-shredders

Scottish island of Mull, where two gerbils were hired as paper-shredders. In no time the little creatures had gnawed secret files and official directives into tiny pieces that could then be safely discarded. "It was a feast for the eye to watch Otto and Shredder at work," says Constable Armour. But there's nothing new under the sun. In the 1970s the well-known Dutch biologist Maarten 't Hart already valued rodents as document-shredders at the university where he worked. Only he didn't use gerbils but the closely related jirds. "All the stencils, memoranda, appendices, structural amendments, policy papers, minutes and committee reports that land on my desk and in my pigeon hole in such vast quantities every day", he wrote, "disappear unread into the jird bin, where they're instantly ground into fine shreds. In this way I can single-handedly undo some of the effects of the university's Administration Reform Act, which has turned the institution into a perpetual talking machine." Reading this makes me nostalgic. Stencils! They still had stencils then! What was perfectly normal only a short while ago – a steady flow of stencils that even an army of rats couldn't keep up with – was stopped almost from one day to the next. But not by the ravages of time. Old techniques haven't died out, they've been pushed aside by new ones. Shredders not jirds now gnaw away at faxes and printouts.

Decay is indestructible. So it's not a product of time, but a measure of it. The measure. The person who sees decay as time's work confuses the clock with the time it tells. Time is measured by decay and decay by time. This explains our morbid fascination with age. We always want to know the age of everything: dinosaurs, Kenny Rogers, Robinson Crusoe, the cock on the weathervane, the dog at our feet. In courtroom proceedings, the name of the suspect is omitted but his age is included; that will do almost as well to help you work out what kind of person you're dealing with. A 37-year-old man? Of course he did it: 37-year-olds can't be trusted. Whether you're a man, a monument or a melody, your age says a lot about you. If you're not a young dog, you're an old fox; he who isn't a fresh bud must be an old goat. There's a reason why it's considered impolite to ask someone their age: by telling your age, you expose yourself. Before you know it, you're stripped bare.

You can make a profession of guessing ages – by becoming an antiques dealer, a matchmaker, a palaeontologist, an auctioneer, an inspector of wares or a train conductor. The fact that you can earn a living doing it means that you can cash in on age. The older an antique vase, the higher the price it commands. Whisky has to be old enough, milk fresh enough. In the art world, the opinion of an expert can mean the difference of millions of pounds. On the basis of paint, style and canvas, such an expert can make a reasonable estimate of the age of a work of art, especially if he's experienced through age himself. But the best experts are forgers. They can make new paintings look old and old fish fresh. Most paintings in museums aren't forgeries, but the age of most food in shops has been seriously tampered with. Thanks to conservation, deep freezes and cold stores, the food on your plate looks much younger than it is. This saves the food industry millions of pounds. In the academic world, forgeries can turn whole world views upside down: one human skull as old as that of a dinosaur, and the whole theory of evolution would be blown to smithereens.

Plants and animals often don't look their age. One is more marked by life than the other. To get around the difference between biological and chronological age, biologists are forever searching for the absolute characteristics.

The age of a tree can be told by counting its annual rings

The example given at school is trees, with their annual rings, although by no means all schools tell you that the tree has to be chopped down first before you can count its rings – which can hardly be the intention where monumental oaks are concerned. So, you have to make do with estimates based on the size of the trunk and the location. Growth rings are also used to establish the age of shells and ear-stones (the calcite and protein deposits in the internal ears of fish). Both accumulate more calcium during summer than winter. If the age of the herring or the cod is known, you can work out how many you're allowed to catch. Biology students are mad about such studies. They count the ear-stones and eat the fish. For a deer's age, look at the branches of its antlers. And to know whether a hare is young or not, pick it up by its two wrist joints: if the forelimbs break, it's young; if they don't, it's old. There's a weak spot in the young bones – the *Stroh'sche Zeichen* (Stroh's nodule) – where growth will still occur. In the case of a gift horse, of course, you don't do this. You're not even allowed to look it in the mouth: you might see the worn-out state of the gift from the worn-down state of the teeth. Still, at the butcher's, you do check carefully to see if the meat is fresh. At the supermarket too the consumer is forever assessing age: is the lettuce still crispy? Are the bananas too brown? Are the apples too wrinkled?

On the basis of all these estimates, counts and measurements, fish quota are established, food shopping is done, horses are traded and tree-felling licences are issued. People cash in on age in the form of retirement pay or old-age pensions. You also need to be a certain age to be eligible to vote, receive housing subsidies or start a restaurant. But how is human age ascertained? People don't have annual rings or ear-stones, and the state of your teeth says more about your dentist than your teeth. In search of criteria, hair colour, lung capacity, ear size, number of wrinkles and many other characteristics have been studied – and rejected. They vary too much. In the same way as a metal ring tells the age of a canary, the birth certificate is the only thing that tells the age of people. But it can easily be forged. If your father registered your birth a year after you were born, you'd never notice that you were a year older than you always thought you were.

In the absence of scientific criteria, people are experts at estimating each other's ages. We spend hours doing it. People watch and are watched all day long. As long as you don't gape, everyone finds it quite normal. And when you're watching television, even that's allowed. The news, a talk show, the umpteenth interview: in terms of image, they're mostly faces (often the same ones too), yet we can sit and watch them night in and night out. Rather than listening to voices on the radio, we look at faces on the television. And what we like most is deciding how old they are. Secretly guessing ages is the most

We love looking at faces and guessing how old they are

popular pastime in trains and waiting rooms. If an audience looks at me intently when I'm giving a lecture, I know it's having difficulty deciding how old I am: wrinkles, drooping eyelids, a head that sinks vulture-like into my shoulders – everything is taken into consideration, even, in the event of further examination, calcification of toenails and loss of potency. Many scales are used for weighing, but even then, the results are often inaccurate, if for no other reason than that biological age doesn't increase nearly as steadily as chronological age. You can't go grey in one night but you can grow old, while a different, more heavenly, night might leave you looking years younger for months on end. But the fact that you can be so far off in your estimate is precisely why it continues to be such fun. And so, tirelessly, the estimating and re-estimating continues and every sign of transience is interpreted in the light of eternity.

The flowering of wrinkles, the withering of flowers, the tearing down of the house you were born in – each in its own right is a clock by which to read the time. The decay of the earth itself is the classic timepiece, with the disappearance of the cliffs of Dover as the second hand, the erosion of the Swiss Alps in the Rhine as the minute hand and the cooling of the earth's magma as the hour hand. The accuracy of this instrument depends on how uniformly its hands move. According to Charles Lyell and James Hutton – the fathers of geology – it was a reliable timekeeper. In their view, the past could only be understood if it was assumed that the same laws were applicable then as now. The present was the key to the past. If you knew how deep the gorge was that a river had gouged out over a ten-year period, you could figure out how long it had taken the river to make the whole gorge – and hence the age of the gorge itself. Similar methods of calculation could be used for thick layers of sedimentary rock. The results were invariably different from what the Bible said. If you added up all the patriarchs with all those who begot others, even with a lot of fiddling, you'd still never end up with an earth that was more than a few thousand years old. If the Bible was the earth's birth certificate, it was a false one.

While geology was busy disproving the Bible, physics was also grappling with the earth's age. In the mid-eighteenth century, the French naturalist George Louis Leclerc, better known as Comte de Buffon, worked on the premise that the earth had once torn loose from the sun, as a flaming ball of liquid. Tests with hot balls of all sizes and materials taught him that the earth had taken 74,047 years to cool down to its present temperature. That age had to be increased by "roughly" 74,832 years because while the earth was cooling it was still being warmed by the sun. Partly because of the incorrect estimation of this last factor, de Buffon was about 60,000 years off. A hundred years later, William Thomson (later Lord Kelvin), was still inaccurate in his estimate when he allowed the earth – on the basis of a 500-million-year-old sun – a few million years at most in which to cool down before life could develop. This created problems for Charles Darwin, who estimated that he needed several hundreds of millions of years for his evolution. Proof for this sea of time was only supplied after he died – when radioactivity was discovered. This revealed that substances disintegrated into other substances with predictable regularity. Instead of millions of years, there now turned out to be hundreds of millions of years available – more than enough to allow seas and continents to change places, fish to crawl onto land, elephants to grow trunks and Darwin to think up his theories of evolution – but far too many to fit into imaginations.

In today's schoolbooks, the history of the earth is presented as a clock according to whose time man only arrives on the stroke of midnight. But there are more striking examples of "deep time". American geologist John McPhee compared the existence of the earth with the old definition of the yard – the distance between the king's nose and his outstretched hand. McPhee claimed the king could have wiped out human history with just one swipe of his nail file. Mark Twain had better images, though poorer figures, on hand:

> Man has been here for 32,000 years. That it took a hundred million years to prepare the world for him is proof that that is what it was done for. I suppose it is, I dunno. If the Eiffel Tower were now representing the world's age, the skin of paint on the pinnacle-knob at its summit would represent man's share of that age; and anybody would perceive that that skin was what the tower was built for. I reckon they would, I dunno.

Although our memories don't go back that far, the mountains of fossils and sedimentary deposits are there to remind us of our prehistory. Without decay, we would never have been able to measure that time. This is amazing. But what is even more amazing about all the ways of measuring age is the conclusion that young can grow old but old can never grow young. We can only remember the past, not the future. Time only moves from the incomprehensible beginning to the inevitable end. However much we'd like it, it can't be reversed. Time is an arrow with a tip that points in only one direction.

But what about trees, fashion and politics? There, everything keeps going round in circles. Leaves fall off and reappear, only to fall off again; skirts get shorter only to get longer again before they shrink for the umpteenth time; and in politics, Labour and the Tories take turns pushing each other out of the same seats. James Hutton realized that rocks couldn't go on eroding forever. If they did, there'd soon be no mountains left. The accumulation of new soil would no longer be able to compensate for what rivers and seas had washed away, and man would starve for lack of arable land. That can never have been God's intention. So the only other possibility was that new mountains and rocks occasionally had to appear, what had worn away occasionally had to be recycled and the earth occasionally had to push its way upwards. Breaking down alternated with building up, and this – according to Hutton – in endless succession. Time didn't travel like an arrow, from creation to ruins, but like a circle, from creation to ruins to creation to ruins to creation. For Hutton, time was a circle that kept repeating itself, whereby you knew which phase of the circle you were in but not which rotation – the thirty-seventh or the nine-hundredth. In this way, Hutton himself, the same man who gave us our sea of time, denies history.

We're not used to this. The Bible raised the Christian world on Biblical history, a story which begins with Creation in a certain number of days, reaches its pinnacle with the resurrection of the Son of God, and ever since has been heading ineluctably towards the happy or unhappy end at the time of the Last Judgment. But even in the Bible, there's "nothing new under the sun":

All the rivers run into the sea; yet the sea is not full; unto the place from whence the rivers come, thither they return again. The thing that hath been, it is that which shall be; and that which is done is that which shall be done: and there is no new thing under the sun. Is there any thing whereof it may be said, See, this is new? It hath been already of old time, which was before us.

According to the Book of Ecclesiastes, this applies especially to the origins of our calendar – the heavenly bodies:

The sun also ariseth, and the sun goeth down, and hasteth to his place where he arose. The wind goeth toward the south, and turneth about unto the north; it whirleth about continually, and the wind returneth again according to his circuits.

There was little that Galileo and Kepler's astronomy could do to change this. Day in and day out, the earth revolves around the sun. At the rate of 270 kilometres a second, this little star races through the Milky Way. Even though the North Star is often not in the north, it comes quite close to it every 26,000 years. But the newest astronomy reveals that after completing one revolution, not a single heavenly body ends up exactly where it began, if for no other reason than that the universe is continually expanding. Just as the New Testament seems to be a repetition of the Old, the world is forever the same yet always different. Time isn't an arrow or a circle; it's a spiral that keeps turning in circles, only to end up somewhere else.

Naturally, people have always tried to compress that spiral and then release it so that it rises up again like a phoenix out of the ashes. If a toy rabbit stops working, you simply put a new battery in it and it merrily starts drumming again. Can't the same be done with a live rabbit or a person? Can't you put new batteries in a person? After all, you don't throw away a transistor radio just because the batteries are finished, do you?

At the beginning of the nineteenth century, shortly after batteries were discovered, man envisaged a bright future in this regard. What Galvani had managed to do with frogs' legs, the Scottish physician Hue thought he could do with a whole person. It was to take place in 1818: a corpse would be brought

Engraving from Galvani's *De Viribus Electricitatis in Motu Musculari*, 1791

to life by electricity. To ensure his wares were fresh, Hue had bought the corpse directly from its original owner, a murderer condemned to hang. The corpse-to-be used the proceeds to brighten up its last living days with gin, beer and beef. An hour after the execution, it lay on the marble dissection table of Glasgow University, surrounded by a crowd of curious scholars and other sensation-seekers. Dr Hue inspected his new acquisition with satisfaction: no welts on the neck, a peaceful expression on the face, everything neatly in place. Next to it stood a row of 64 containers full of gently boiling chemicals. A wire led to the neck of the hanged man, whose spinal cord had been deftly exposed. A switch was turned on, the current did its work and the whole body began to shudder, as if life were doing its best to reclaim it; the public shuddered too. The current was turned off and the body became a corpse again.

An electrode was then attached to a leg that was held by an assistant. The power was turned on, the leg straightened and the assistant fell over backwards, kicked by the dead man, who apparently wanted to be left alone. They tried other organs. Driven by the current, the whole chest began to move so that you would have sworn it was breathing. The deathlook, influenced by the effect of the electrodes on the facial muscles, flickered between a grimace of despair and expressions of blind fury, sadness and recrimination. Irately, the dead eyes glared out into the room; accusingly, an electrically driven index finger pointed to the bystanders one by one. Spectators walked away, one fainted, the seeds were planted for chronically recurring nightmares. But despite the best efforts there in Glasgow, the motor wouldn't start; no new life could be breathed into the corpse. Dead stayed dead. Legally, it would have been more interesting if they'd succeeded, for what do you do with a hanged man who comes to life again? If death can be undone, the death sentence becomes a mere outing and loses its point.

Bringing the dead back to life: today it is the stuff of horror films. And rightly so. Nothing is more frightening than when something that looked dead suddenly comes alive again. The only problem is that it can't happen. A dead person coming to life is like a stone falling upwards, pieces of china jumping up off the ground to become a teapot again, or bullets returning, in

Today the dead are only brought to life in horror films

succession, to the barrel of the pistol from which they came. If you see something like this on television, you know the tape is being run backwards. But if you can rewind a film, why can't you rewind time? Why can't ruins decay into a temple, an old man grow up into a baby, a tree return to its seed? What's so special about time that it's irreversible? In essence, nothing. Past and future are equal in nature's eyes. Other important variables in physics can be reversed without difficulty: pressure can be increased and decreased at will; volumes can be expanded and contracted; atoms that have just arrived at B will happily return to A. There's no reason at all why the sun shouldn't move from west to east, and the North Pole automatically become the South Pole if the current of an electromagnet is reversed. What disallows stones from falling upwards and teapots from spontaneously reconstructing themselves is the second law of thermodynamics. This is the law which says that everything was better in the past. The past: when there was order and regularity.

"Thermodynamics" sounds more modern than it is. It's the science of heat and mechanical work and it has its origins in the era of steam and coal. According to its first law, energy can't be created. What power stations do is to convert energy that was already in the fuel into electrical energy. Energy can be changed from one form into another. According to the second law, however, some of the energy is lost, in the form of heat, during the process. There are always atoms that won't allow themselves to be confined by the new straitjacket of electrical current or mechanical movement and that, drifting randomly, escape the new order. Thus, after every conversion,

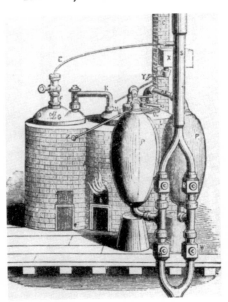

Energy conversion in Thomas Savery's
steam machine, 1702

there's less energy left at the end. A car produces less propulsion energy than is put into it in the form of petrol energy. A lion gets only one-tenth of the energy that its prey, the antelope, accumulated during a lifetime of eating grass, and that energy in turn is only one-tenth of the energy originally converted by the grass. Nine-tenths of something as orderly as an antelope evaporates into disorderly warmth. All around you, disorder is being created out of order. The term for disorder in physics is "entropy". According to the second law of thermodynamics, this always increases in spontaneous processes. That can't be a good thing. The world is irreversibly heading for a "heat death". This book is contributing to the process. To write it, I stoked hundreds of sandwiches and glasses of gin into my intestines. The tiny bit of order that this book contributes to the thinking about decay is no match for the disorder that the burning of such delightfully ordered material as sandwiches and gin brings with it. This is why one animal ages very differently from the next. Everything is organized – up to and including reproduction – according to an orderly system; after that, it's every animal for itself. It doesn't matter how it meets its end, as long as it does. So, does spontaneous order never emerge from disorder? Of course it does. Of course a bag of Scrabble letters can fall to the ground in such a way that they form a meaningful text. However, the chances that a meaningless jumble will emerge are much greater. Of course there are particles of air that go from low to high pressure. But so many more particles do the opposite that it's safe to say that in normal life, air goes from high pressure to low pressure until the pressure is equal everywhere.

While engineers a hundred years ago were as pleased as Punch with their thermodynamics, biologists were tearing their hair out. If order tends towards disorder, how can anything as super-orderly as life ever emerge from disorderly matter? The answer has everything to do with the definition of entropy. The second law of thermodynamics applies only to closed systems that are in equilibrium. If there's one system that's not closed and not in equilibrium, it's the earth. The earth gets bombarded with sun energy. Although plants absorb only a small percentage of that energy, it's still enough to build whole forests and steppes, and to allow all of those millions of animals with their well-organized bodies to drop, lay and bear new animals. People build up their own complex bodies by breaking down the complex bodies of cattle and cauliflowers. The energy that remains is used to build – or destroy, depending on our mood – cities or world empires. Because we can't do this fast enough using only our own energy, we also tap the energy sealed in coal and oil. What the earth has managed to squeeze out of the sun over a period of hundreds of millions of years, man succeeds in racing through in a few centuries. Decay doesn't have to be slow.

Order only moves in one direction: from more to less. This is why time is also a one-way street. The passing of time is always measured according to this decline in order: a compressed coil returns to its original shape, grains of sand flow downwards, atoms disintegrate. Time follows the principle of entropy in just the same way as one arrow follows another. Because we can't reverse time's arrow, there's nothing left for us to do but make that arrow as long as possible: we do everything within our power to extend our earthly life. People who do so successfully are viewed with admiration. Birthdays aren't mourned, they're celebrated. The older the people around us become, the greater the chance we too will gain a year or two. Even in a country as highly technologically developed as the Netherlands, it's still common to fall back on incantations to invoke this: "*Lang zul je leven!*" (Long may you live!), we say to each other on birthdays. It helps; all Dutch people are convinced of this. Why else, year in and year out, do they continue to arrange a special day so that they can receive these wishes? Once a year, for an entire day, in exchange for little snacks and drinks, guests wish you a long life. Dutch people are even prepared to do something on these occasions that they would never dare do otherwise: sing in public. "*Lang zul je leven*", pipes Auntie Dot, "*in de gloria, in de gloria*" (In the glory, In the glory), wherever that may be. The efficacy of this unscientific way of prolonging life was already refuted scientifically in 1872 by Francis Galton. In his *Statistical Inquiries into the Efficacy of Prayer*, he investigated whether all those congratulations to the Queen and all those prayers for good health on behalf of the other members of the royal family paid off. Despite the many blessings of "Long Live the Queen!", it turned out that the 97 persons of royal blood whom he studied, with their average 64 years of age, generally lived three to five years less than all the vicars, men of law, doctors, officers and merchants for whom there had been much less praying and much less flag-waving.

So what is it that makes us live to be 100? That's the question aspiring young journalists are expected to investigate when they're sent out to attend birthday celebrations of centenarians. Dutch writer Godfried Bomans wrote an inimitable sketch about this:

> "Is father home?" I asked the old-timer who opened the door.
>
> He nodded, and showed me into the little room where an even older old-timer was sitting, who was almost dead. I quickly grabbed an ear trumpet from the wall and shouted into his ear: "Many happy returns!"
>
> "You're mistaken," said the old old-timer in a lacklustre voice. "Father is upstairs."

Resentfully, the centenarian interrupts his exercises on the gymnastic rings to answer the question about how he came to be so old:

> "It just happened. Every year you grow a year older, it's in the nature of things. It's a question of patience. When I turned seventy, I was still an insignificant little man, my time was still to come. My eightieth birthday was even an all-time low; nobody understood why I didn't die. But then I gradually started coming into my own. When I turned ninety, people started pointing at me, and by the time I was ninety-five, all my problems were behind me."

This explanation is remarkably accurate. Every time a person turns a year older, they're more likely to die soon. All life insurance agents since Benjamin Gompertz know that an adult's chances of dying double every seven years. Statistically speaking, escaping death quickly becomes an impossibility. But God is merciful towards the oldest. After the age of 80, the risk of dying increases less sharply each year. A 105-year-old has a greater chance of becoming 106 than a 104-year-old does of becoming 105. By the time you reach this age group, anyone with even the slightest inclination to die has already done so. The struggle is now between the very fittest. To grow extremely old, you first have to be old, and to be old, you have to have the aptitude. Old age is inherited. If both your parents reached more than 70, your chances of reaching 90 or 100 are twice the normal rate. Nine out of ten nonagenarians or centenarians have at least one parent who was older than 70.

To the chagrin of politicians needing babies to kiss, the number of centenarians is increasing. In the Netherlands alone, there are more than 1,000 over-a-hundreds. The over-a-hundreds are the fastest-growing age group. This doesn't have much effect on the maximum age of people in general, though. Despite the fact that an increasing number of people are living longer, man as a species isn't. Centenarians were around long before the Christian calendar, and if you look up the ages of famous men and women in the encyclopedia, you'll also find that many in the distant past reached ages that a modern old people's home would be proud of. How is this possible? Everything is getting better all the time, isn't it? And aren't we jumping a centimetre further or higher or both every year, as well as running a second faster? Yes, we are. But as a species man isn't living one second longer. Most records are being improved upon simply because more and more people are participating in the competitions. A hundred years ago almost no one took part in sports. (Women are still clambering to close the gap.) And in many poor countries, people have other things on their minds than who can throw the discus the furthest. The only competition in which everyone, everywhere, always participates is the competition of who'll live the longest. With the exception of a few suicides, no one, ever, anywhere, wants to die, and as a result, the record for longevity has long been firmly established. Despite all the advances of medical science there's no touch after the finishing touch.

The winners of the "who'll live the longest" competition seldom look like winners. This is because it's not so much a case of winning as of not having lost yet. Nor can the winners enjoy holding the record for long. Until recently,

the oldest person alive was Jeanne
Calment. On 17 October, 1995, she
broke the previous record held by
Japanese Schigechiyo Izumi, who
died in 1986 at the age of 120 years
and 237 days. Although most of
the claims for record ages are
made by Russians or other remote
peoples, Jeanne Calment was just
an ordinary Frenchwoman. She met
Vincent van Gogh when she was

To the chagrin of many politicians, the number
of centenarians is increasing

young. On her 120th birthday, Madame Calment said that she had had to
wait 110 years for fame and that now she intended to take advantage of it.
"I'm waiting for death and for journalists," she said. Although the man
who bought her Paris apartment counted on being able to live there after
what was her undoubtedly imminent death, in fact he died years before she
did. Jeanne Calment passed away in 1997 at the age of 123.

If animals were allowed to participate in the Olympic Games, humans
would cut a sorry figure. Whales grow larger, cheetahs run faster and eagles
see better. At nature's Olympic Games, we would be the odd-ones-out with
our oversized heads. But we'd make a pretty good showing when it came to
longevity. Among mammals, we're even the champions, having a considerable
headstart on number two, the elephants. Although hunters like to brag that
elephants can reach more than 100 years of age, the record is, in fact, 70. That
was the age at which No. 1342, better known as Kyaw Thee, died in Burma,
where it had earned its keep by dragging tree trunks for man. This is a
good example of the theory that domestic animals have a better chance of
living longer. Horses are next in line after elephants. Old Billy is supposed to
have been the oldest. He died at the age of 62, finally retired, after a life of
pulling boats along English rivers and canals; that was on 27 November, 1882.
His stuffed head can still be admired in Bedford Museum. Dogs don't even
come near that age. They turn up their paws at the age of 29, seven years earlier

Old Billy died in 1882, at the age of 62

than the oldest cats. If you really want your family pet to outlive you, you'd better choose a parrot instead of a horse. A record of 73 years has been registered for the greater sulphur-crested cockatoo (*Cacatua galerita*). Incidentally, birds grow much older than you'd think. If they don't die from hunger or mishap, they hardly wear out at all. Classic proof can be found in the photographs of the Scottish ornithologist George Dunnet, together with the same ringed fulmar. In one photograph Dunnet is clearly 40 years older than in the other – taken in 1950 – but you see almost no changes at all in the bird after all those years. Only its ring is a little worn.

There are also old-timers among fish. A famous example is the pike that was caught in Kaiserwag Lake at Württemberg, Germany, in 1497. According to the copper ring around its neck, it had been released into the lake in 1230. The skeleton of this monstrous fish – which was therefore supposed to be 267 years old – is still preserved in Mannheim Cathedral. Examination by a nineteenth-century anatomist revealed that the bones were from several different fish. Astronomical ages are also claimed for the koi fish, a fancy carp in Japan that can cost a fortune. The specimen called "Hanako", which belonged to the headman of Higashi Shirikawa village, is said to have reached 223 years of age. The growth rings apparently proved it. But a koi can develop two or three of these rings every year. The same applies to whales, which, according to the grooves in the baleen, should be able to reach 90 but in reality only make it to half that age. Reliable ages for fish stop at about 80. Grandfather sterlet, an ordinary sturgeon (*Acipenser ruthenus*), for example, died after an aquarium life of 69 years and eight months in the Amsterdam Zoo; ages of 80 have been reported for lake sturgeons (*Acipenser fulvescens*) from Lake Winnebago in Wisconsin. With ages of 50 or 60, apes come nowhere near our record, but they still belong to the top league of long-lasting mammals. Not because they're related to us, but because they're so big. In every group of animals, it's the giants that grow the oldest. People are also giants, not only in, but also outside of, their animal group. Although elephants are heavier and giraffes taller, more than 99 per cent of animals are smaller than man.

Man grows as old as he is large. He lives at a leisurely pace. It's a question of physics. Just as a tower clock is more sluggish than a wristwatch, an elephant

Elephants in the zoo can live to be 70

is more sluggish than a mouse. Big elephant legs and little mouse legs are governed by the same laws as a clock's pendulum: the longer the pendulum, the slower it swings. Even if elephants could patter around like mice, they'd soon become overheated and expire in a cloud of steam. Relatively speaking, they have little skin through which to expel heat from their enormous bodies. This is a question of geometry. For the sake of discussion, imagine an elephant as a cube. If it's 25 times larger than a cube-shaped mouse, it has 25×25 or 625 times more skin, but no less than $25 \times 25 \times 25$ or 15,625 times more mass. It has to take it easy or it'll become as hot and as breathless as a fat man on a staircase in summer.

People move more than elephants, but all the scurrying and scuttering and squealing of small fry like tits and mice is too much for us. We giants are driven mad by midgets like mosquitoes and flies that we can't catch. People and mosquitoes live in the same space but a different time. We seem as slow to mosquitoes as the hands of a clock seem to us. If mosquitoes are as fast as lightning, we're as slow as molasses. To follow a bluebottle, you have to film it with a high-speed camera and play the film back at our tempo. Only then can you see how its wings, during each of their lightning-quick beats, turn on their longitudinal axis, and how the wave-like vibrations pass from the base of the wings to their tips. Conversely, to understand us, mosquitoes would have to watch a slow-speed film of the kind in which we see flowers shoot up, unfurl, produce seeds and wilt, all in 20 seconds.

What's the point of living a life as long as ours if it passes so slowly? If you grow ten times as old by living ten times as slowly, in the end you experience nothing more than you would have in a short, fast life. Elephants and mice are given the same "life-film", but one is played ten times faster than the other. In its few years, a mouse has put one foot ahead of the other, been frightened and loved, as often as an elephant has in its many years. Each probably feels it has lived just as long as the other – a lifetime.

Even in the same human life, human time isn't always the same as clock time. If you stand and wait until your tea water boils, every minute takes an hour; if you're trying to finish an exam, time is up before you know it.

According to Diderot, "work has the advantage that it shortens days and lengthens life." For a couple who are in love, time seems to stop altogether – but not for long. Warmth has almost the same effect on time as love does; a clock depends on the temperature. When hot, a pendulum gets longer, the glass of an hourglass expands and chemical processes take place more quickly. The same applies to our biological clock. American biologist Hudson Hoagland realized this after he'd gone to get medicine for his wife, who lay in bed with a temperature. She blamed him for taking so long, even though he'd only been out for 15 minutes. Because she had a temperature, her bodily processes took place more quickly and, as a result, time seemed to pass more slowly. This is particularly useful to know when you're dealing with old people, who invariably complain that the final years of their life rush past much faster than the first years. Because life's processes slow down as we get older, time speeds up. We should really set fires under those old folk. Perhaps that's why it's so hot in old people's homes that visitors wilt away, together with the flowers; the only things that thrive there are old people and tropical plants. But time seems to pass faster for older people than for children when it's cold too. While to a child, a week is a year and a year an eternity, to an adult years tick away like weeks and eternity is just around the corner. There's a good psychological explanation for this: we measure our lives in terms of the life we've had, of the life behind us. For a child, a year is one quarter of its conscious existence; for an old person, it's no more than 1 per cent. Gone, just like that. But there are compensations. The older we get, the more time we spend dozing. While young people are always busy doing things, old people don't mind just sitting. And just being. Until they cease to be anymore.

They say people are walking clocks. There's a clock in each of us that tells us when to be sleepy and when to be hungry. Fiddling with that clock leads to jet lag or bonuses for working the night shift. Animals even follow the clock in their love life: they're impotent for most of the year. But we're not walking clocks, we're walking clock shops. Each part of our body, each organ, each cell and each chemical reaction has its own tempo, its own time, as in a shop full of clocks, each of which ticks through the same 24 hours but each with its own hands, its own nervousness, its own ticking. The tempo varies but the number of ticks remains the same. Whether you're an elephant or a mouse, after roughly 1,000 million ticks your heart stops ticking. A mouse races through its ration in a few years, an elephant takes 50. A mouse has a nut-sized heart, a cat a mouse-sized heart, a lion a cat-sized heart and a whale a lion-sized heart. But, during their respective lifetimes, all of these hearts, whether big or little, pump about the same amount of blood into each gram of body. Thus, during its lifetime, a gram of mouse converts the same amount of energy as a gram of elephant: approximately 650 kilojoules. Once the cells in that gram have used their allotment of energy, they die. Bird cells convert three to four times more energy before they depart this life. How they manage to do this is still a mystery, but it suits the birds just fine. With their little bodies and active lifestyles, they need more energy per gram than mammals do.

Amphibians and reptiles lead a low-key life. They're cold-blooded, which means their body warmth comes primarily from the environment. While a mammal produces just enough energy per kilogram to light a flashlight (four watts), a kilogram of crocodile only converts one-tenth of that amount. This is one reason why crocodiles are less delicate than mammals of the same size. A Mississippi alligator that had been brought to the Dresden Zoo in 1880 lived through the Hitler era. Tortoises grow even older than humans. The records are 138 years for a common box tortoise and 152 years for a Marion's tortoise. Among other cold-blooded creatures, some lobsters, salamanders and termites live to be at least 50. Of the termites, it's the queen who lives the longest and continues to lay eggs throughout her life.

Although you can't prolong your life, you can try to manipulate time. If a year had 100 days instead of 365, you could suddenly turn 365 years old instead of 100. But no one would be fooled. The rising and setting of the sun would quickly have you running off to the clockmaker, and even in the deepest, darkest corners of a mine your biological clock would keep you on the straight and narrow. One day of fiddling away time in a plane is enough to make your body protest. Even stranger is the fact that our government thinks it can fool us with Summer Time, which, to make matters worse, is introduced in spring. Suddenly everything happens at the same time, only an hour earlier. Suddenly an hour is stolen from the night and given to the day. And we only give it back in October. But it's not that easy. He who tries to fool time with his clocks is like the fat person who fiddles with the scales. The inventor of Summer Time comes from the same stock as the person who keeps flying westwards to escape getting old. Fiddling. Due to Summer Time, we do everything earlier because we think it's later – and not just during summer, but all year round. In the Netherlands it's Summer Time all year round. In mid-winter, the sun isn't directly overhead at twelve noon – as you learned at school – but only at a quarter to one. In Germany, the sun is on time, but that's where our time comes from anyway, so it's still always Summer Time in the Netherlands. During the summer it only gets worse. Then you have to wait until a quarter to two before the sun shows up for its twelve o'clock appointment.

All this is done to keep the pubs, camp sites, caravans, canteens and snack bars happy. The longer there's light, the more money we spend. We have to "recreate" until we're ready to drop. We're surprised when we look at our watches on summer evenings. Isn't our leisure time finished yet? Sighing, we go on enjoying ourselves. After all, Summer Time is good for the environment, isn't it? That remains to be seen. Ultimately, the environment benefits about as much from having its own time as it does from having its own minister. The energy you save by using less artificial light gets used up right away by the evening's recreation. If you want to save energy, you shouldn't be fooling around with time, but with temperature. You'd be surprised at how quickly consumption would drop if water were declared boiling at 80 degrees Celsius instead of 100.

So why do we prefer fiddling with time? Not because we want light longer, but because we want dark shorter. People are afraid of the dark. It's the end of the day and you never know if there will be another one. Sunrise comforts, sunset does not. Sunset; that's what we call old people's homes – to scare the life out of the oldies right from the start. There's no question that their days are numbered. Wouldn't they love to gain an extra day or two!

It is possible. You can easily extend a human life by four years, just by adding an hour and a quarter to each day. Just as you call three o'clock "two o'clock" once a year, when age starts to gnaw at you, you could simply call 34 "30" once in your life. Henceforth, you would just add on the birthdays from that 30 onwards. When you're 50, you'd actually be 54; when you're 66, you'd actually be 70. No one would notice, because everyone already acts four years younger than they are anyway. Thanks to cosmetics and sports, no one looks as old as his father or mother did at the same age. If everyone above the age of 34 took four years off his age, the balance between number and behaviour would be restored. That would be good for all those everyones, and also good for the State. There would suddenly be many fewer pensioners! The only problem is that you wouldn't reach the same grand old age as before. But something can be done about that. In the same way as you give back to Winter Time the hour you stole during the summer, everyone could give back those four lost years when they die. Die young and still grow old, it's yours if you want it. It may be wrong to fool yourself, but it's not difficult.

You do it every year. On your birthday you celebrate the fact that you've grown a year older. But, in fact, that's only partly true. The skin you have now, for example, wasn't there a year ago. Your skin has been replaced 15 times during the past year, the soles of your feet even more often. Your red blood cells would no longer recognize their confrères from the preceding year because they only live for 100 days, and your intestine wall is re-upholstered every three days. Hair and nails grow out of your body before your very eyes. You're a ragbag of old and new parts, like a pair of endlessly patched trousers. Actually, each part of your body should celebrate its own "yearday" or, for those parts that don't make it that far, "monthday" or "hourday". On average, by the time you're 50, you're about as old as a toddler. Aside from a few skeletal muscles, only your nervous system and you actually have something to celebrate on your birthday. Your nervous system preserves the memory of long-lost parts. What we call dying is only the death of the leftover parts; almost everything else was jettisoned long ago without a single tear being shed. That last little bit is allowed to stay alive just a bit longer so that it can be amazed by time.

I turn and turn the glass as hours go by,
as if to prove what I well comprehend;
I seem to see the *how*, but it's the *why*
that causes every hair to stand on end.

What if I can, or cannot in this way
come slightly closer to the puzzle's core?
I've seen the loveliest tresses turn to grey;
the loveliest lips will one day speak no more.

Time's cancer rules, pervading all we see,
And so at bottom every last sensation,
when set against this dire *fait accompli*,

lacks relevance, is just contemptible.
I'm curious to know what revelation
Can make this travesty acceptable.

JEAN PIERRE RAWIE

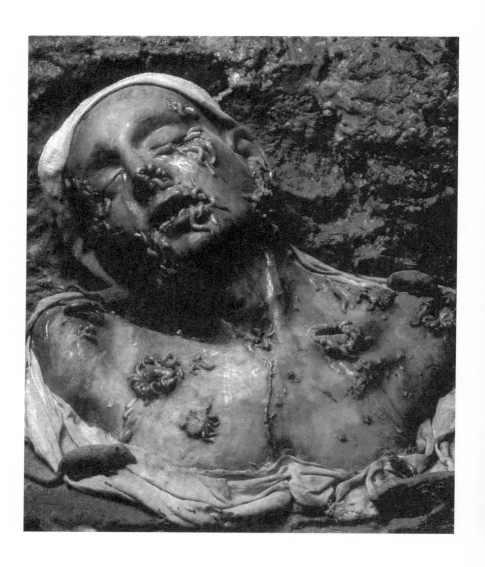

7

ASHES AND DUST

It doesn't all end with death. Dying only gets into full swing after death. Your body faces its biggest challenge then. That whole structure of bones and muscles and nerves and mucus has to return to dust. Only then can the Biblical prophecy be fulfilled. Ashes to ashes, dust to dust. Everything built up during a long lifetime – first inside the womb, then outside it – has to be broken down, into the building blocks of the building blocks. What shortly after death looks as it did just before death – as if someone forgot to turn it on – has to disintegrate to become the soil in your garden, to blow until it becomes the wind in your hair. A hellish task. Dying happens automatically, but passing away, like the locomotives in the locomotive graveyard: how do you do that? How do you spirit away a whole body? You wouldn't even know what the answer is alive, let alone dead. Yet you will do it. Like all those who have done it before you.

Passing away is a self-accelerating process. At first, you still see the tree clearly in the fallen forest giant; it's just the world that looks a bit tilted. Later, when the leaves refuse to reappear, it becomes more difficult to deny death. Gradually, the tree turns into wood, the wood into rot and the rot into mould. The same process, at a quicker rate, can be observed in the fruit bowl, where leftover fruit does away with itself in a cloud of mould. Reputations also spoil quickly. And dead baby birds. I know an illustrator who prefers to draw them dead rather than alive. "Dead baby birds are more lively," he says.

If you die, you become something: you're suddenly a corpse. Your skin goes pale and limp, your muscles slacken, your pulse disappears. The body

A dead barn owl

cools down at the rate of one or two degrees an hour, until after about twelve hours it really feels like a corpse. Doctors have increasingly complex criteria for establishing when death has occurred – brain waves, chances of reanimation, the demand for organs – but the rest of us instinctively know the moment a person becomes a corpse: it's when it's scariest. Everybody becomes scary when they die. As long as you're alive, even your worst enemy occasionally shakes your hand. But who would dare shake hands with a corpse? Who would lovingly run his fingers through a dead man's hair? That the hair of the living is already dead doesn't make the hair of the dead any less scary. There's always that lingering fear that something is still alive in the dead person. You just never know if there isn't a live spirit hiding somewhere in there that forgot to vacate the body. It could be the devil or a disease-causing organism, but whatever it is, it's still scary. Just how dead is a doornail?

Whispering, people pass the coffin; silently, they stand at the graveside, afraid to disturb the dead man in his eternal sleep. Imagine if he were to wake up! Although we no longer carry corpses around the house to chase away evil spirits, no corpse goes into its grave before we've closed its eyes. Somehow we manage to look at it, but it must promise not to look back. And, after all that being looked at, it must go into its coffin post-haste. The ancient Greeks had a beautiful word for coffin: sarcophagus. *Sarx*, from the word "sarcoma", means "flesh" in Greek; *phagein* means "to eat". A sarcophagus is a flesh-eater. So get on with your meal!

Coffins don't eat flesh. To ensure that His creations are destroyed on time, the Creator created an enormous demolition factory. Worms and maggots, vultures and crows – they can't wait until we're dead. A person dies like an empire: at the hands of barbarians. As soon as the last lines of defence have fallen, they storm in and smash everything to pieces. No longer thwarted by life forces, they wade through breaches in the hated skin like marines through holes in the Atlantikwall or like woodworms through a church roof abandoned by both God and The Heritage Trust. As befits barbarians, they don't run off with all the pomp and splendour that made them so green with envy; the grandeur has to be destroyed, pulled down, razed to the ground. It's not so much the materials as the structures that are the first to go. Deprived of cohesion, the empire then disintegrates of its own accord – from ashes to dust. It decomposes.

To give you some idea of how eager your fellow creatures are to wipe out all traces of your existence, just put a dead mouse in the garden on a beautiful summer's day where the cat can't get at it. In no time, the mouse starts wasting away with such a fury that it almost seems to come to life again. It's a hive of activity inside there, as if there's a party going on. The little legs move as if they might run away at any moment. But it remains a sack race; the commotion is caused by maggots eating away at the insides. They appear so quickly that it's almost as if they fell out of the sky. In a way, they did, too. They emerged from the larvae that flies laid in the cadaver. Long before we smell anything, they've been attracted by the earliest whiffs of rotting gases, such as skatole, acetone and acetic acid. And they have to get there quickly because they never know when such an opportunity will arise again. Carrion doesn't walk away or fight back but it is delivered very irregularly. To increase its chances of reproduction, the grey flesh fly (*Sarcophaga carnaria*) lays ready-made larvae instead of eggs. Within a week or two, they've eaten themselves sick and for a while it becomes quiet inside the mouse. The larvae are pupating. In the end, the mouse takes to the air – in the form of hundreds of fat flies.

As thanks for its stay on earth, the mouse leaves its skin behind, a gift for the skin-lovers and hair-lovers. These species are rarer than carnivores, because skin and hair are full of tough materials like collagen and keratin. But you're better acquainted with them than you realize. Since it makes no difference to them whether we use skin or hair to make leather jackets, fur coats or woollen sweaters, you can also find them in your clothes cupboards. Fur beetles (*Attagenus*), carpet beetles (*Anthrenus*) and other bugs think a house is one big, wonderful animal graveyard. Of the moths, only the common clothes moth (*Tineola bisselliella*), originally a tropical creature, is found indoors, while the case-bearing clothes moth (*Tineola pellionella*) is found both inside and outside the house. It's outside that you can see where moths developed their unusual predilection for something as inedible as keratin: they're found mostly in birds' nests.

Birds are major users of keratin. They use it to make feathers. But although they've been doing it for more than 100 million years, their feathers still aren't

Above and opposite: Medieval
danses macabres

very good. They have to be replaced every year. During the moulting season, enormous quantities of worn-out feathers are discarded. Based on an average of 5,000 feathers per bird and a bird fauna of 50 million specimens, 250,000 million feathers end up in the Dutch environment each year. Without moths and bugs, the country would have long since become one big eiderdown. Each year anew, insects prevent this spectre, risking their lives in the process. As an insect, it's extremely dangerous to live in the nest of insect-eaters. So, thousands of years ago, it was quite shrewd of them to decide to move into man's houses: most people don't like to eat insects, and, until only a short while ago, pesticides had never been heard of.

It's not nice to watch moths and beetles consuming our clothes and carpets, but this is only a mild taste of what's in store for us later. They like human skin and hair too. By the time the keratin teams show up, armies of other eater-uppers are usually already hard at work. In the same way as a cowpat hosts hundreds of different creepy-crawlies – one in pursuit of the juicy soft parts, the other mad about the bits with something hard in them that it can sink its teeth into – each stage of our body's decay attracts different species. Using the classic work *La faune des cadavres*, by J. P. Mégnin, the police can even establish how long a body has been dead, in what season it was murdered and whether it was dragged. To cover his tracks, a modern murderer almost needs to be an entomologist.

Initially, in addition to flesh flies, the smell of fresh corpse attracts hordes of wasps. Although not specialized in corpse-eating, omnivores like wasps quite enjoy a mouthful of corpse with their lemonade. Later, the burying beetles (*Necrophorus*) arrive. They're often up to their ears in mites (*Poecilochirus necrophori*), which eat the larvae of the flesh flies instead of the flesh. Out of gratitude for the free ride that the mites were given to the corpse, they thus kill the burying beetles' rivals. Beetles and wasps are nearly always accompanied by ants. There are many more kilograms of ants on earth than there are kilograms of mammals or birds walking or flying about. Together with tropical termites, ants clean up three-quarters of all small corpses.

After a few months, when the fat becomes rancid enough, bacon beetles (*Dermestes*) and bacon moths (*Aglossa*) are attracted by the volatile fatty acids. Once the proteins become cheesy, the cheese skippers (*Piophila casei*) appear, to do the same thing in old corpses as they do in old cheese. The drier the cadaver becomes, the greater the number of cadaver beetles (*Hister*) that show up to keep the burying beetles company. Dessert is for the spider beetles (*Ptinus*), which live off the excrement that the earlier corpse visitors left spread over the bones like a grey film.

Some insects can't wait until the corpse becomes corpse-like. The greenbottle (Lucilia caesar) lays its eggs in open sores. In those little islands of death in still-live bodies, the maggots live off the dead flesh. They used to be bred for the purpose of keeping wounds clean. Today, they're important mostly for the information they can give to forensic entomologists. If greenbottles are found in fresh corpses, then they must have found an artificial opening through which to enter the body; there's probably a gun or knife wound somewhere. Conversely, beetles found in corpses that are actually too old for that particular species suggest poisoning – with parathion or arsenic. These poisons delay the development of flies and beetles, while cocaine would help an insect develop more quickly.

In Sweden, in the spring of 1728, Carl Linnaeus was bitten in the arm by a "worm" while gathering plants. His arm became so severely swollen that a doctor had to cut it open from top to bottom. After "the most blood-curdling pain" and "intense aching and throbbing", Linnaeus was saved. Out of revenge, he described the worm as Furia infernalis: the infernal fury. The creature has never been heard of since; today, it's thought to have been the larva of a flesh fly. When the time came, however, Furia had little opportunity to take a bite out of its nomenclator. Before he died, Linnaeus gave instructions for his funeral. "Lay me in the coffin", he wrote, "unshaven, unwashed, undressed, wrapped only in a sheet. Nail the coffin shut right away, so that no one will see what I look like."

You don't need a coffin to keep out insects. Fine-meshed netting is equally good. In the past, everyone had net-covered boxes, like my grandmother's, for storing meat. But such boxes don't prevent decomposition. No netting is fine enough to keep out bacteria. Even if there was a box with such netting, it wouldn't help, because the beef steaks and pork chops were already teeming with bacteria at the butcher's. Bacteria come from the air, the butcher's fingers or the paper the meat is wrapped in to keep it hygienic. They don't multiply very quickly, because there isn't much for them to eat as long as the cells are intact. This situation changes radically the moment the meat is cooked. The cells burst open, thereby allowing access to the nutrients inside. At the same time, the bacteria are killed. Problems arise most often when the meat is only seared. If it goes back into the meat safe at that point, a veritable feast breaks out – an orgy – during which the bacteria multiply at breakneck speed. At room temperature they double every half-hour. After about ten hours, 100 bacteria have become 100 million bacteria. Compared with the mass of the piece of meat, the mass of all of those minuscule organisms is negligible. You don't taste them at all. What gives them away is their lack of toilet training: one by one they urinate into your food. This gives it a strange flavour. The food is – we say – spoilt; the food was – they say – delicious. But they've only done their duty. How could they have known that the butcher had a monopoly on this meat and that it wasn't a dead bird waiting to be cleared away in the name of public health?

Bacteria are too small to be seen by the naked eye. Sometimes, though, you see them at work before you smell them. The trick is to attract the right species. The best way to do this is with fish: bacteria love fresh plaice and cool sea water. Put the plaice into a deep dish half-covered with salt water (30 grams of salt for every litre of water) and store it at five to ten degrees Celsius. Go and look at the fish occasionally in the dark of night, and after about 24 hours you'll be surprised: it's luminescent! It radiates a mysterious glow, especially near the edges, where the air and water meet. In reality, of course, it's the bacteria that are doing the glowing. Bacteria glow the way some people make noise when they eat. But not for long. The sickly smell that greets you at fishing ports tells you that there are many more phases of rotting yet to come. New species of bacteria take over the demolition process, followed by other species and yet other species – until there's nothing left of the plaice.

We, too, are alive with bacteria. After we die, they can settle down to do their work. In about two or three days, our body is covered with a thin green layer of bacteria, most of which originated from our intestines. The only humans to whom this doesn't happen are newborn babies. A baby is born sterile. It gets intestinal bacteria from, of all places, its bathwater. Stillborn babies decompose poorly due to lack of bacteria. Sometimes they even mummify.

One form of decay is as distasteful as another. People don't like the idea

Children's mummies in the catacombs at Palermo, Sicily

of decaying like a rotten fish or an abandoned piece of cake. Fortunately, nature offers an alternative to being slowly gnawed at or licked on by an endless procession of increasingly sinister organisms. More than half of all cadavers get swallowed in one bite by large carrion-eaters, like hyenas or vultures. Where once there were cadavers, within no time there's only carcasses. Europeans have made more frequent use of this method than you'd think. Not with hyenas, but with sharks. In the past, many a young European died at sea and was thrown overboard. Sometimes the whole crew was served to the sharks, ship and all. Generally, the boat withstood the talons of time better under water than on land, where oxygen always aggressively penetrates every corner, causing things to rust and supplying the woodworm with air to breathe.

On some South Sea Islands, the dead are also entrusted to sharks, and in India the Parsees leave their loved ones in open spaces where carrion vultures can get at them easily. Europeans are offended by such things. The thought of being eaten by a carrion-eater reduces us to carrion. And carrion-eaters already have such a bad reputation. Those men in zoos who stand looking

In India some Parsees leave their corpses for the vultures

only at the vultures, hyenas and rats are viewed suspiciously. Vultures are seen as shifty scavengers, with their necks worn away from rooting around in steaming carcasses. It's a good thing there are no vultures in our part of the world. But we have our own corpse-eaters – eels. Eels are caught using half-rotten horse heads, their long bodies wriggle out of the exposed eye sockets. Show that on television and the price of eels would drop for weeks. "Slippery bastards", anglers call them, "thieves in the night". Dutiful little housewives wrap these *Unterfische* in newspapers so that later, at home, they can chop them, live, into little bits.

Thanks to butchers, poulterers and fishmongers, we know what dead cows, chickens and cod look like. But what does a dead swan look like? When did you last see a dead squirrel? Do swallows take their dead south with them? Of course, you occasionally see a dead animal in the woods or near the water, but this is nothing compared with the dying that's constantly taking place around us. All those little birds born in all those nests, all those flies swarming round your head on a warm summer evening: they must die to create space. All those shells on the beach have fought their own personal death-struggle. There's a lot more dying than living in nature. Every year 50 million birds are wasted in the Netherlands; 50 million last little breaths – that's quite a stiff breeze. It's a wonder it doesn't rain dead birds when we go for walks in the forest. Paths should be impassable because of dead frogs and finches; your mouth and nose should be stuffed from the torrent of dead insects. If animals were to groan during their last hour, the noise would be enough to waken the dead.

In reality, what you see are mostly victims of the only predator that doesn't eat its prey: the car. The victims pile up along the roadside. The hedgehogs are the LPs, the frogs the CDs. Because the hedgehogs roll up into little balls when they hear the wheels approaching, the transition from three- to two-dimensional seems more dramatic, but flat is flat. Frozen in their jump, their little feet poignantly extended, you can imagine how they've been hurrying to reproduce so as to stay ahead of our death-thirsty driving. In total, several million mammals and birds are killed by cars annually in northern Europe. A working group in the Netherlands once counted 125 species of birds and 24 species of mammals – which really should be 125 and 25, because man is also a mammal.

It's not a pleasant sight – all those big and little bodies along the road-side. But it's not objective always to side with the dead. For carrion-eaters like crows, jays and gulls, the motorways are tables that have been laid and

Insects on the body of a Highland cow

rush-hour is dinner time. Ravens have lately been released in many places in Europe. They used to live off the remains that wolves left of their prey. Despite valiant efforts, wolves haven't reappeared yet, but cars make up for a lot. No efforts are spared to ensure that car wheels, and thus ravens, have enough food. Ecologically correct grass banks attract birds, illuminated signs above motorways attract rabbits, and we help frogs across the road every spring to ensure there'll be enough creatures to flatten on the return journey in the autumn. You only see signs of "Game Crossing" far from where game really crosses the road. Have you ever seen a deer near such a sign? For carrion-eaters, the road is a banquet of hedgehog cookies and mashed frog. Nature-lovers can trade in their bird-watching excursions for good old road side tourism.

However clearly biology teachers might explain the usefulness of carrion-eaters to the ecosystem, we prefer to keep our human remains out of this cycle. Consequently, most of our dead are buried. We say we'll never forget them and then quickly put them out of sight. Let bygones be bygones. This custom may originally have come into existence to prevent the enemy from walking off with our bodies and eating them. Today, the war dead are still buried, so that later, when peace breaks out, they can be accurately counted, to determine which side really won the war.

But whichever way you look at it, for reasons of hygiene, burying is a wise move. Germs can do little damage underground and there's no air to spread the stench. The condition is, of course, that the corpses must be buried on time. In northern climates, this usually means within five days. But that isn't always possible. During the Starvation Winter of 1944–1945, when the Germans had taken most of the food for their own troops and temperatures were severe, so many people in Amsterdam died that the graveyards couldn't keep up with it. At one time, there were 150 bodies in the Zuiderkerk waiting to be buried. Due to the shortage of wood for coffins, special coffins with trap-door bottoms were built. At funerals, the trap-doors were opened, the bodies slipped into the grave and the coffins were ready for re-use. Thanks to the extremely cold temperatures, no diseases broke out. Elsewhere in the city, the District Undertaker's Office once found a dead woman in bed in a one-room flat "who had died at least five weeks earlier. The children and husband slept in the same bed and lived and played in the same room as the deceased lay. All this so as not to lose the dead woman's food coupons." If it had been summer, the body would long have been visited by flies and bored through by maggots. The chances that this would happen in winter or in the freezers of modern mortuaries are much slimmer. But their day will come.

According to his "A Contribution to the Study of the Grave", in the 150 bodies he exhumed, M. G. Motter found a complete fauna of insects, worms and other small fry that had managed to find their way through the cracks into the coffins. Reconciliation with the idea of death is asking too much, but I find it comforting to know that when the time comes, many beautiful beetles and flies will be born out of my old man's body. According to Motter, people-eaters aren't fussy: men get eaten by the same species as women and old corpses by the same species as young. The state of buried corpses says more about the composition of the soil, the quality of the coffin and the level of the groundwater than the variety of the fauna says about the corpse. A buried corpse is of little use to forensic entomologists. The guests

Shortages during the Starvation Winter of 1944–1945 led to the re-use of coffins in the Netherlands

arrive at the party in random order and they all arrive too late. Underground, a corpse takes eight times as long to decompose as it would above ground and four times as long as it would in water. It can take about ten years before a grave is more or less empty. But often, after 20 or 30 years, there's still a disconcerting amount of grandad left underground. The worms would certainly have *wanted* to reach him during that time, but they would have been unable to. Chipboard coffins, synthetic body bags and plasticized veneer blocked their route to the dinner table. Such substances don't decay so much as saponify – which doesn't make the work of the gravecleaners any more pleasant.

The advocates of cremation prefer to dissolve into thin air. A corpse decomposes 100,000 times faster in the flames of a crematorium than through gradual incineration in the little digestive organs of all those little body-eaters. At 1,000 degrees Celsius, a corpse is turned to ashes and dust within a few hours. There's no faster way of making the Biblical prophecy come true. But the Church resisted cremation for hundreds of years. In the Christian tradition, cremation was reserved for atheists, witches and sodomites. To banish the evil of their deeds from the face of the earth forever, the perpetrators themselves had to go up in smoke. Their ashes and dust were scattered far and wide by the wind, so that at the Resurrection their bodies would never be able to pull themselves together again. Enemies of the Church knew this and preferred martyrs to be burned at the stake, which, of course, made martyrdom all the more martyrly, and thus strengthened the Church.

Until well into the twentieth century, voluntary cremation was as uncommon between the two European rivers, the Meuse and the Rhine, as it has always been common along the Ganges. Between the two World Wars, a fierce struggle was fought along the Thames to remove the legal impediments to cremation. Arguments for and against flew back and forth. Printers prospered thanks to the countless folders, brochures, pamphlets and what later came to be known – in all innocence – as "propaganda". The cremationists eagerly pointed out how the gnawing of worms and the rooting away of bacteria disturbed the everlasting peace in the grave from the outset:

To be sure, deceptive poetry can be concealed in it – if on a June day, full of the scent and sheen of summer, we gently lay our "beautiful" dead in warm, dry graves under the cooling foliage, where birds sing of the triumph of life; of course, it is possible that, at that moment, through the field of tombstones and crosses, we don't see the horridness of the funereal onslaught of wriggling graveyard creatures. But how does it feel when we have to move our beloved from the warmth of the home and the happiness of the family to a cold, wet, muddy, solitary grave – on a cold, sad, wintry day?

The fear the living feel for the diseases of the dead was also widely exploited:

In how many graveyards does the groundwater have access to the coffins, and who's to say whether infection isn't carried over long distances via unknown underground waterways?

Nor were the arguments of the opponents always very sound. Cremation was said to deprive the ground of important substances. In particular, too few corpses might lead to a shortage of ammonia in the ground. The cremationists were accused of selfishness and materialistic cold-heartedness in this respect. One Dutch professor even maintained that extra bodies should be buried in infertile areas. To help exploit wasteland, he had four major heath cemeteries in mind for the Netherlands. And, just as the cremationists talked of "the gruesome process in the grave", the advocates of burial gladly told of how, in the infernal oven, corpses resisted the hot air "by writhing, wincing,

A nineteenth-century oven for cremating bodies

moving, opening their eyes, flailing their hands, and so forth". It's the "and so forth" that still sets my imagination alight. But, according to Dr L. A. Rademaker, chief executive of the Board of the Dutch Voluntary Cremation Association, there was nothing to worry about. Together with a small group, he had witnessed the serene reality of what took place in the ovens with his own eyes:

> The event moved us deeply, but it was an emotion of great beauty. The afternoon was well advanced and dusk was already falling on that late autumn afternoon when the oven fell open and suddenly gave us a view of a horizon that glowed as intensely as on a beautiful summer evening.

Poetically, at least, cremation is a match for burial. The only real difference is speed, because the end-product is – and always will be – the same. Dust. Almost three kilograms of dust. It's a sobering thought to realize that in the end, the greatest poet, the cruellest dictator, the most beautiful woman will all be reduced to little piles of dust. But there is some consolation: the biggest trees, the most poisonous snakes, the most spectacular mountain

Propaganda for the legalization of cremation, c. 1880. The captions read: (top) "Two kinds of interment"; (below left) "Piety personified"; (below right) "Horror itself"

ranges will also be reduced to dust. The dust you walk on, the dust you inhale and the dust you see dancing in the sunlight.

Everything and everybody turns to dust. Regardless of what it's made of, as long as it's small enough to float around in the air, dust is dust. By definition, dust particles vary in size from one-millionth of a millimetre to one millimetre. What you see in the sunlight at home are mostly cotton, wool and paper fibres. On the floor it's mostly sand and clay particles carried in from outside. But, of the particles smaller than one-tenth of a millimetre, more than half come from us. More than 50 per cent of the finest household dust is made up of flakes of skin. If you think you're losing yourself, just look on top of the cupboard: run a finger across it and you'll be standing face to face with the state in which you'll find yourself in only a few decades' time. One slap of your hand on the mattress and you'll be flurrying about in the air.

Household dust hosts an entire ecosystem. A few fleas, of course, and doubtless some lice, but mostly mites, mites that live off mites and mites

that eat the excrement of the mites that live off mites. In the dust you return to, you're spared the company of this small fry. There's nothing edible left in the three kilograms of dust in the urn or in the grave under the long-since-crooked gravestone. These small fry don't like flames any more than worms or bugs do. The only thing left in your dust is inorganic material – indigestible, suitable only for consumption by plants. Miracle Gro. Phostrogen.

Photomicrograph (x150) of household dust mites (*Dermatophagoides oteronyssinus*)

What kind of dust is it that we return to? What's in a person? It's common knowledge that we consist mostly of water. You could make 66 pots of tea or coffee with the water from one body. After we die, that water evaporates. Dying is mostly evaporating. What remains consists primarily of nitrogen and carbon, absorbed from the air for us by plants. The chimneys of the crematoria and the body openings of the little body-eaters put most of these gases back into the atmosphere, where they intensify the greenhouse effect. In the few kilograms of dust that are left, there's enough phosphorus for 50 boxes of matches, iron for one solid-iron nail, lime for whitewashing a chicken coop, and just enough potassium for one roll of pistol caps. Altogether, this would cost you no more than a few notes at the chemist's, but you'd need a few extra for accessories. Most people haven't been just people for a very long time: aside from the lead and mercury in our fillings, a growing number of bodies also accommodate hips made of nickel and pacemakers containing platinum. Many a corpse is no more than chemical waste. Burying and cremating are both bad for the environment – because the ash from cremating has to be disposed of as well. And we already have enough environmental problems without the accessories. According to one Dutch ecologist, "cremation ash, given its ecological effects, is a calcium-phosphate fertilizer polluted with heavy metals." The phosphate fertilizes directly, and the calcium indirectly, by releasing nitrogen from the soil, so that you can clearly see where grandad's ashes were scattered in the forest – near the stinging nettles. The same nitrogen-loving plants thrive in graveyards, where, according to one environmental encyclopedia, they form a heavy burden on the soil due to the enormous concentration of organic material that has to be absorbed. "In a few cases, a higher concentration of nitrates and a higher consumption of oxygen was observed in the ground-water." To bury is to create compost.

More important than what's in a person is what's not. If you analyse the ashes and vapours into which we disintegrate, it turns out that we're not even remotely made of "the dust of the ground". The same applies to "every beast of the earth" and "every fowl of the air", yes, even to all the "grass" and "tree-yielding fruit". Of the 92 elements from which "the dust of the ground" is made, only 16 are found in all living things. Species that especially want to be noticed can choose from an additional eight elements, but then the pie's finished. Three-quarters of the elements in nature are not used by its inhabitants. What's more, life absorbs its building blocks in very different proportions to those in which the building blocks are present in the environment. Hence, there's much more phosphorus in you than around you.

The biggest difference between the dust on your mantelpiece and the dust in your urn is the silicon content. While the dust on the floor of your house is about 30 per cent sand, clay and other silicon-rich materials, none of the silicon around you is in you. Varieties of grass arm themselves with silicon to combat grass-eaters. A few plants and animals absorb silicon, but they make shells from it. Yet the silicon shells of little siliceous creatures have as little to do with their living tissue as a scooter helmet does with the driver of the scooter. Although silicon constitutes more than one-quarter of the earth's crust, it's unthinkable that we would ever take a bite of it. Nothing is more disconcerting than when our teeth grind as we eat a sandwich at the seaside. Mortar-layers, housewreckers and concrete drillers know why: fine particles of silicon nestle so deeply in their lungs that they become short of breath and can eventually die from it. Construction workers hack and drill their way to their own death.

Their sacrifice is not in vain. Because so few organisms like to eat silicon, we can make things from it that will stay with us for a long time: houses, roads, tombstones. Nothing is so inhospitable to an organism as a bare stone: no nourishment, cold one moment, burning hot the next. In the end, even the hardest rocks split, but that's due more to acids than to the assault of time. Many decomposer-organisms don't mind putting their teeth to the grind, but chewing boulders takes too long. That's why the tombstone is usually the last part of a grave to go. If it contains a lot of silicon, even the inscriptions will still be legible. It's no coincidence that the oldest texts are preserved in granite columns and clay tablets. The oldest commandments were engraved in stone tablets.And what do we do? Entrust our texts to paper!

Paper is rich in plants. This makes it "yummy, yummy" for the thousands of organisms that live off dead plants. We're grateful for their presence in compost heaps and on forest floors, but librarians dislike book-lovers that read with their teeth. Bookworms know no better than that books are there for them to eat, marinated for them over the years. And where did all those gramophone records go? When I was a boy, millions of 78s were sold every year. For "Jingle Bells" and "White Christmas" alone, millions of records must have been pressed. But few people know what they were made of: shellac. What's shellac? It's a preparation made from insect pap. A 78 consists of many thousands of ground bug corpses. With some records, it's as if you can still hear them whining. Shellac comes from an Indian species of scale insect (*Laccifer lacca*), with a shell made of a useful kind of resin which is incorporated into shoe polish, furniture wax, playing cards, pool balls and even insecticides. But the peak years were still the 1920s and 1930s, when the whole record industry depended on it. In 1927 and 1928 alone, England, Germany and France needed 20,000 tons of shellac for their 250 million records. Thousands of millions of little bugs gave their lives for it.

With the exception of the odd collectors' item, all those records are now broken, gone or lost – banned from our lives as symbols of the brevity of Top Ten fame. Yet "Jingle Bells" and "Rudolph the Red-Nosed Reindeer" still manage to find their way to us every year. The 78s were copied: first onto singles and EPs, then onto LPs and cassettes, and now onto CDs. Who knows what new wrappings they'll be kept in next? But with every record we break, some music is lost. That's what makes us so angry. It's not the shellac that matters so much; it's the grooves. By damaging the shellac, you also damage its contents. When records break or books fall apart, it confronts us with the reality of the essence of life: the knowledge that with the decay of the body, the spirit is also lost. What churches try to compensate for with faith, biologists do with reproduction. All bodies are eventually broken down or eaten up by worms, but the spirit can easily be reproduced: in reprints, progeny, "The Best of . . ." CDs, microfiches, or even better – thoughts.

The spirit can be saved, but what happens to the body once it's gone the way of all flesh? If a record that was never played still has music in its grooves and a book that was never read is still full of exciting words, isn't there still some information left in a corpse? How exciting is a corpse? For some people, it's the most exciting thing there is. A necrophile is aroused by dead bodies; he strokes their cold cheeks and has sex with the rotten flesh. Because a corpse is dead, necrophilia falls by definition under fetishism, but it's fetishism of an exceptional kind. Even from an era as perverse as antiquity, only a few examples of necrophilia are known. Herodotus mentions the tyrant Periander, who had sex with his wife after she died, and also casts a shrill light on the Egyptian mummy-makers:

> Women who are very beautiful or highly respected are not handed over for embalming right away, but only after three or four days. This is to prevent the embalmers from having sex with them. For it is said that someone was caught having sex with the corpse of a recently deceased woman; his colleague had betrayed him.

A necrophile needn't have an affinity with gerontophilia. In 1996 a Belgian man was arrested who had an obsession with dead children. The first time his house was searched, the Belgian gendarmerie found pictures of a nine-year-old girl who had been run over a year before somewhere else in the country. The photographs had been taken in the hospital. Later, it turned out that the 59-year-old E. M. had in his possession, not only masses of child pornography but thousands of photographs of a total of 1,630 different children's bodies. According to one of the Flemish dailies, he had been scouring graveyards and mortuaries for photographs for about 40 years:

> Often he posed as a member of the young victim's family. Apparently he went into action as soon as the burial was over. That's why it was almost never noticed that the earth had been dug up and the child's grave desecrated. It is thought the necrophile was active throughout the country. Since the bachelor only travelled by public transport, it has been ruled out that he also engaged in child kidnapping. E. M., from Etterbeek, had been running a shop selling religious objects on the Kammenstraat since 1993. In any case, the loner was

a strange person. He was also fascinated by death, which is revealed by the fact that he slept in a coffin.

Grave desecration carries a one-year jail sentence in Belgium. However dead a corpse may be, it's against the law to touch one. Even though the living spirit has officially left it, the next-of-kin don't want to take any risks with the mortal remains. While vicars and priests mind the spirit, gravediggers and pathologists exorcize the body with disinfecting substances. Ordinary people console themselves with music, and listen to Frank Sinatra. There's even comfort in that. Because, what's comfort? Comfort is knowing it could always be worse. You could be saying your farewells with Engelbert Humperdinck's "There Goes My Everything!".

8

SOUVENIRS

Your eyes are for looking, your ears are for hearing, your nose is for remembering. It takes only a whiff of something and you're suddenly back 10, 20, 50 years. When I smell aniseed, I'm helping my grandmother in the kitchen again. Let Brussels sprouts cook in the pan and my Aunt Anna looms up out of the steam. If I sniff on a pot of glue for just one second too long, I'm back glueing my home-made cut-out elves onto a piece of yellow cardboard, almost a lifetime ago. That's a time machine hanging down over your lips.

Want to return to your early school years? Just sharpen a pencil. The smell of the shavings immediately reminds you of Mr Blackthorn, the Battle of Hastings, the girl with the plaits. And what does a pencil sharpener cost? Almost everything you need for a trip through time can be bought for a few pennies at the corner shop. Your university years? Let a glass of beer go flat, preferably next to an overflowing ashtray, and you'll experience the morning after the night before all over again. You don't need to have children in order to relive your youth. You can still bake shortbread without them. If the smell of Christmas baking makes you long just a bit too much for your childhood, open a bottle of real cod liver oil: you'll be cured in no time.

A child's hand holding an eye. Eighteenth-century anatomical specimen by Frederik Ruysch

You refresh your mind with your nose. What you revive with it are mostly memories. Unfortunately, there are no little machines on the market to help you have those smells on hand at all times, so most people make do with photographs for their memories. They cut pieces out of the world and make it flat. Then they glue these pieces into an album, in memory of themselves. But it never amounts to much. An album with 600 photographs taken at an average shutter speed of 1/60 of a second barely captures ten seconds of your life. Whole silver mines are converted into film and seas of fixer swill through the world's darkrooms just to give every citizen his ten seconds of

THREE CONTINENTS IN SIX WEEKS

Most people make do with photographs to revive their memories permanence. Laden with cameras, holidaymakers don't see the sites because they're too busy taking pictures. People are dogs in reverse: while dogs urinate on trees so they can leave something behind, people snap photographs so they can take something with them. The main activity of these excursions is to capture a photographic scent. What else is there to do? A boring day and a picture of it: that's the essence of day-trips. Such a photograph only becomes meaningful later, when reality is no longer within reach and only the memory remains. This kind of later is always too late.

A photograph is a tangible memory. That's why it's wrong to paste photographs into albums. Nothing makes you feel better than loose photographs floating around in an old shoe-box – groping around in your memory, rummaging through it until you come across something you no longer knew you knew. People and animals long since dead come alive in you again. The most exciting pictures are those where the dead are standing next to the still living; it creates a bond. But with photographs, it's still a question of making do. Many millions of pounds of snapshots and family portraits are stored away – unlooked-at – in cupboards and attics, until they finally disintegrate back into amorphous silver dust. Forgotten memories. Forgotten because they can only be looked at, not smelt. Photographs can't recreate smells. And shapes alone don't take you back far.

Eyes are impassive sensors. Observers. What they see doesn't affect them. They merely gather information, then pass it on to sophisticated brain centres, where it is analysed and converted into images. Smelling, on the other hand, you do yourself. You smell with your brain. Your nose doesn't serve your brain – it's part of it. Moreover, the centres of smell, like the parts of your brain responsible for hearing, are much older than the centres of sight. They're linked – and almost no intelligence is needed for this – to feelings. That's why you can move someone more easily, more directly, with music than with a beautiful building or an ugly painting. That's why smells have an immediate effect on your mood. It's difficult to describe a smell other than by comparing it with a familiar smell, but the emotions it evokes have stirred the most florid language from poets since time immemorial.

Left: A model of the Temple of Minerva Medica, cork on wood, first half of the eighteenth century
Right: The Eiffel Tower as a pepper-mill

Without eyes you have less of a present; without a nose you have less of a past. There's a great need for the past. The more past you have, the longer you live; a person is as old as his oldest memory. Every lost memory shortens life by the duration of what's been forgotten. Certainly, there's a lot of the past, but for it to be useful, it needs to be brought to the present. Since it's impossible to pour the ocean of the past into the bucket of the present, we consume the past through random sampling. A song by the Rolling Stones brings your teenage years back to life, a glass of flat beer recalls your university days, and the words "*Arbeit macht frei*" are all it takes to throw you back to the Second World War. Play the Stones or take a sip of that beer and you relive moments from the past, as if you can fiddle extra minutes out of time. This need for what one Dutch photographer calls "the evasion of transience" is so great that we intentionally collect objects now, so that later, when now will be then, we'll be able to think about now again, and experience that now, again. Such objects are called souvenirs. A souvenir can be a photograph or a miniature Big Ben, an admission ticket or an old magazine, as long as it has a place and, especially – a time. It's a coordinate of time and space, part of our identity, one of eternity's straws to clutch at.

What's characteristic of a souvenir is that when it's bought, it isn't yet a memento. It still has to become one. Whether it will be a good memento remains to be seen. So buy your holiday souvenirs carefully; don't lose sight of the fact that they'll later have to remind you of something, they must have a special air about them. A little Eiffel Tower is useless. It takes you back, not to France, but to souvenir shops, not to Frenchmen, but to tourists. Buy a bottle of Pernod instead. One whiff and it's as if you have a beret on your head, because you'd never drink the stuff at home. From a walk in the forest, take home a piece of moss. From the beach, a salty piece of meerschaum. Because at the seaside, too, it's smells that arouse the deepest nostalgia. Nowhere is the sea more beautiful than in the dunes, where you can't yet see it but where you can just catch its scent. That scent is both the anticipation and the memory – suddenly so crystal-clear – of all those times when you walked through the dunes on your way to the sea: the little boy with the spade, the blissful half of a young couple-in-love, the father with his first child. Without smells there is no past; without a past there is no present. Life in all its richness can't be seen – only smelt, inhaled.

We aren't the only ones who long for the smell of the past. Salmon travel thousands of kilometres through the sea in search of the rivers where they were born. The smell of their roots is irresistible. Neither currents nor locks nor waterfalls can deter them in their attempts to spawn in the place they were conceived. The fact that they remember this smell all their lives can only surprise those people who've never again breathed in the smell of the family boatyard, the caves visited during school outings, the coal-bin beside the furnace.

Smells disappear. Not even a salmon thinks the Clyde smells like it used to. And where can my nose find the fumes of an old Morris Minor? Not in a car museum: that only smells of polishing wax. Which canal still dares reek of canal? How exciting it must have been in the days of the steam engine, when, *en route* to the station to start your holiday, you could already smell the steam from kilometres away! What's the point of continuing to visit the sick if hospitals no longer smell like hospitals, or of visiting old-age homes if they no longer smell of old age? How, in this age of the extractor fan, are young people supposed to acquire the smells they'll need later – so that they can remember them later?

If institutions for the sick and the elderly already look like offices, how will we remember them? Locomotives look less and less like locomotives, cars less and less like cars, Rolls-Royces less and less like Rolls-Royces. In time, this will be a problem. But one thing will remain the same: souvenirs will remain souvenirs. It's easy to distinguish utensils from souvenirs on the mantelpiece. Animals on the mantelpiece are nearly always souvenirs: porcelain cocker spaniels, stuffed squirrels, cuckoos in cuckoo clocks. They refer to the origin of the souvenir. When a man goes somewhere, he's supposed to come back with something. Preferably a dead animal. He hangs those pieces that weren't thrown up in the air or made off with, above the mantelpiece as a memento of the process of destroying. Such pieces are called trophies. But fishermen, too, have their milestones. Fish that are too large to be eaten or thrown back are stuffed. To become a good trophy, an animal has to meet two criteria: it mustn't be dangerous (a tiger-skin rug that still bites isn't a trophy yet), but it must look dangerous. There's more honour in stuffing a pike than a goldfish.

A "toad" purse

A hunter would be ashamed to hang up a trophy he hadn't bagged with his own two hands, but modern tourists aren't troubled by this at all. European customs officers have their hands full with confiscating zebra skins, stuffed mambas, dried sea horses and other remnants of animals killed just so that people in faraway countries can be reminded of the wilds they once braved. There almost isn't enough wildlife to go around. But the natives have a solution for this: they either let the foreign hunters pay dearly to shoot the wildlife or they start their own breeding farm. At last, something to make up for all those little beads and trinkets! The mountains of junk are enough to make one's stomach turn: handbags made from snake skin, egg-spoons from cow horn, table lamps from puffer fish, umbrella stands from elephant legs. You can also do it in reverse – make animals out of other materials: a wooden snake, an ivory turtle, a plastic frog. Sometimes the two are even combined: a cow horn made from real cow horn, an elephant made from real ivory, a baby crocodile made from real, grown-up crocodile. Third-worlders are crazy about this. It saves them the trouble of having to make them into handbags or spoons. It's the measurements that tell you that a souvenir's involved: the stuffed crocodile is a baby, otherwise it wouldn't fit into your hand luggage; the elephant made from real elephant is a miniature. Since we're no longer allowed to wear coats made from the fur of baby seals, play seals are made from it instead. This tendency to miniaturize has made such inroads that many people refuse to believe that a bonsai is the actual tree.

Souvenirs have all the features of genuine kitsch. Instead of being something, they refer to something; they're ugly. And if you can do anything with them at all, it's always something different than what you thought you could do with them: a windmill that mills pepper not wind; wooden shoes that house geraniums instead of feet; cats that turn out to be teapots. A tiny Eiffel Tower that spans an ashtray literally has no prospects. And that's precisely the intention. If you ever went into a lamp shop to buy a wheel with the intention of putting it onto a cart, they'd think you were mad. When you buy a souvenir, instead of choosing something you could use – like a French corkscrew, a Turkish coffee pot or a Belgian gun – you're supposed to buy something utterly useless that gets in the way so much that not a day passes without your thinking of the country it came from. Even from the longest journey man has ever made – to the moon – all they could think of to bring back was a bag of rocks. Souvenirs are useful precisely because of their uselessness. If anyone ought to know that, it's you. Your body is full of souvenirs.

Many of your organs are just hangers-on. They no longer serve any purpose or at least they no longer serve it well. That would seem to contradict everything that the Church has always to' 's about the Creator or that the World

Wildlife Fund has ever promised us about evolution. A Creator doesn't produce shoddy work and natural selection ensures that every organism is optimally adapted. Certainly, evolution ensures that species adapt to meet the needs of their time, but during the course of all the reconstructions and reorganizations that are needed to achieve this, old junk sometimes gets left behind in the body. As long as it doesn't obstruct the body's functions enough to justify disposing of it, it kicks around as if in a teenager's room. Your body may eventually pay a high price for such evolutionary scars, but they're as much a treasure-house for the biologist as a cesspool is for the archaeologist. Every organ that serves no purpose must be a memory of times past.

The classic memory organ – the nose – is itself a memory. It stems from the time, millions of years ago, when we were respectable mammals. A respectable mammal doesn't look with its eyes, but with its nose. A dog couldn't care less how ugly you are, as long as your underwear smells good. His world is etched in smells not colours. As a result, it looks completely different. Take your dog to the library and it won't understand a word of all those books, however exciting the adventures they describe. Conversely, if your dog takes you to the woods, you won't have the faintest notion why it gets so excited near certain tree trunks and why it sticks its nose into something as disgusting as turds. But dogs read just as well as we do. The difference is that they read smells. To a dog, a turd is a true call of nature, a message from its male or female maker. A forest full of turds is a library, with here and there, in the form of an exquisitely peaked piece of excrement, a poem. For dogs, even love is dictated by the nose. If male humans were to behave like male dogs, they would eagerly crowd around recently used women's toilets, stick their noses up women's skirts and pass around soiled knickers instead of Playboy. In reality, male humans love with their eyes. Women are devoured by them so bodies and knickers have to be beautiful. As far as noses are concerned, we're not mammals but birds. Birds smell as poorly as we do. That's why we love them so much. And that's why the association for the protection of birds has a thousand times more members than the association for the protection of mammals. Bird species are colour-coded so that we can look them up more easily in picture books. A guide to mammals, on the other hand, looks like a catalogue for packing paper, with lots of brown and grey. There are no guides to smells.

People look with their eyes. They've been doing this since primate memory. Apes and monkeys swing from branch to branch through the trees. If they used their noses to do this, they might succeed a couple of times, but after that they'd definitely be lying on the ground. A tree-jumper has to be able to estimate its leap. For this purpose, its eyes moved to the front of its face. In normal mammals, like cows, the eyes are on either side of the head. This way, they can see danger approaching from any direction – but they cannot estimate leaps. That's why there are so few cows swinging through the trees. Primates see stereoscopically. Because each eye sees a bit differently, they see depth. But they had to pay a high price to do this. When their eyes moved to the front of their face, there was little room left for their nose. It came under pressure and thus became smaller, until there was just enough of it left for eyeglasses to rest on. That's why we see what we should have been able to smell. Noses, like appendixes, are disappearing. They're rudimentary organs, although one may be less rudimentary than the other.

A nineteenth-century impression of the missing link

Incidentally, such variation is characteristic of rudimentary organs. Since the process of disappearing is still in full swing, we have rudimentary organs in every phase of diminution.

The best souvenir from our monkey days hangs above our buttocks: our tailbone. Touch it and try to imagine it's still a tail. What you couldn't do with a tail! In a park, for example, on a bench in the moonlight, your lover beside you, with a tail as well as two hands. You've still got the muscles for it: there are still little tail muscles on your tailbone. You could wag it if you were happy – if only the tailbone hadn't become so rigidly deformed. There are similar muscles on your ears. Ears are also organs past their prime. Compared with those trumpets on dogs' heads, our ears are toys. You can hear a bit with them, but that's about all. To illustrate the evolution of ears graphically, take a plastic cup. That's what our forefathers' ears once looked like. Put the cup on your left hand and hit it flat with your right. What's left resembles how your ears look today. Those ears have the same muscles that allow a dog to turn its ears so that it can hear cross-eared. There are even people who can still move their ears with them. But they don't do it to hear better, only to spoil family get-togethers. These muscles are nothing more than souvenirs from better days. Like faded portraits on the wall, monkey ears, wisdom teeth and those little snatches of hair in the strangest corners of our body remind us of the forefathers without whom we wouldn't be here.

Deeply embedded in the blubber of whales, somewhere above the anus, lie a few bones that aren't attached to anything. These are the rudiments of a pelvis, reminders of days long past when the forefathers of whales had four legs and were landlubbers. Closer to home, on the dinner plate, in the bumpy protrusions on either side of the chicken, the layman recognizes the wings it must have flown with when it was a real bird. Pick the meat off the bones and you'll even see that those wings were once legs. To find out how, you have to turn to fossils. On the odd occasion, rudiments turn up alive and kicking in real life. Such a reversion to an ancestral state is known as an atavism. A classic example are splint bones in horses. Like us, horses once had feet full of toes. The hooves they walk on today were once the middle toes, the thumb and little finger have disappeared completely and the index and ring fingers have shrivelled up to form the splint bones, which haven't touched the ground for many years now. On rare occasions, though, mares demonstrate the theory of evolution by giving birth to foals with three hooves on each leg.

Human equivalents are found at the circus – men totally covered in hair and women with extra breasts the way you'd expect to find them on sows. The

Italian craniometrist Cesare Lombroso considered criminality just such an atavism. A criminal – but also a whore – was simply someone who had reverted to an earlier, less civilized phase of human development. He thought it could be seen too: a rapist had bulging eyes and thick lips, burglars had deformed skulls and high cheekbones. Followers of Lombroso, among whom was the Dutch psychiatrist Gerbrandus Jelgersma, saw those same auricles that were typical of criminals in the ape house at the Amsterdam Zoo. This nonsense is now a good century old, but it still resonates in words like "mongol". During the nineteenth century, many white people thought negroes and mongols were the missing link between the ape-man and themselves; they might even turn white one day but that was still a long way off. In the meantime, they couldn't even think properly. The English physician John Langdon-Down recognized the backwardness of other races in the insane white people at the institution where he was director.

Followers of Lombroso used this craniometer to measure skulls

As well as mongols, he identified Ethiopians and Malays among his atavistic patients. These last categories have now disappeared from psychiatry but the term "mongol" was used for Down's syndrome sufferers until well into the twentieth century.

Your body is a museum of yourself. It houses not only your origins but also your life story. In the case of those with Down's syndrome, something went wrong with the museum when it was being built, something with chromosomes. Other children are born with spina bifida or webs between their toes. In these cases the error occurred at a later stage: the spinal cord didn't close properly or the cells between the toes didn't die off in time. But even if everything goes as it should, you're still full of scars at birth. There's a scar on your heart that shows where the embryonal passage closed to make independent breathing possible. In girls there's a poorly finished weld – the hymen. The best proof of life before birth is your navel. It's the best souvenir of nine months of warm security. This is why painters of Bible scenes used to have such difficulty with it. Did or didn't Adam have a navel? And why have a navel if you don't have a mother? Many painters managed to avoid the whole touchy business by using cleverly draped vines, which could be made to drape in a more southerly direction on the body if necessary.

After the navel, many other scars follow, both inside and outside the body: wrinkles, clogged arteries, trivia, habits, bald spots and, naturally, memories. The body is the most important museum piece in the museum that everyone makes of the house he lives in. Every object testifies to your existence. Floor lamps, books, bottle openers, dirty socks, stains on the wallpaper – they all reflect your life as well as photographs and letters do. Among them are the milestones: the report cards, the silver-plated baby brush, the tree you planted yourself, important dates and, of course, smells. My late colleague Dick Hillenius considered his garden especially to be "a kind of extended memory". His wild daffodils took him back to journeys he had made in southern Italy many years before. Many of his friends "lived on" in his garden even after they had died, because plants or trees they had given him grew there. And for many years now, a whole garden has been reminding us of Dick Hillenius. It's a strange experience to walk through a garden that clearly used to belong to someone else. I once visited Charles Darwin's garden, the Sand-Walk, where he used to walk in circles every day at noon. There, behind his country house at Down, thoughts were born, arguments were refined and counter-arguments refuted, and, thanks to the theory of evolution, biology finally transcended the phase of astrology and weather reports. Everything was still in place in his study too, as if he might walk in at any moment: the writing plank over the armrests of the chair, the microscope on the windowsill catching the diffuse northerly light, the faithful spittoon poised on the floor. Everything breathed the man. The house was like a sealed memory.

Massive structures have been built in the name of man's collective memory.

The architecture of the British Museum, the Bibliothèque Nationale, the Rijksmuseum in Amsterdam and the National Archives in Washington signify the importance attached to that memory. We think it doesn't matter very much whether Sheridan's horse or John Glenn's space capsule are preserved. But it's invaluable that 60 million such objects are stored in the Smithsonian Institution in Washington. Without a collective memory, humanity is irredeemably lost. Unlike animal society, human society needs more than genes for passing on information from one generation to the next. We also like to pass on what was learned in the course of our lifetime and from previous generations. Museums and archives help us do that. Their thick walls and imposing façades form a heavily reinforced brainpan designed to keep time's destruction at bay, even though museum directors, more than anyone else, know what a futile attempt this is.

Many museums derive their charm precisely from the fact that they fail in this attempt. This is because it's in such museums, where objects are allowed to continue with their decaying, that the past becomes most tangible. The older something looks, the older we think it is. One doesn't judge age by the date on the label; age is a function of decay. They're still around, museums that are also museums of themselves: Sir John Soane's Museum, left for posterity exactly as it was in Soane's lifetime, the Pitt Rivers Museum in Oxford, with its cluttered cases and original small handwritten labels, and the unsurpassed Teylers Museum in Haarlem, with its megalomaniac electrostatic generator, where you almost expect to have occasion to say "Mr Frankenstein, I presume?" at any moment. But they're becoming rarer. Institutions whose job it is to stop time are becoming increasingly afraid of not keeping up with their times. Old objects are increasingly being exhibited in new accommodations. The old collection of horrors that used to belong to the nineteenth-century Dutch physician Willem Vrolik, with its Siamese twins in formaldehyde, has been moved from the dimly lit Anatomical Embryological Laboratory in the old centre of Amsterdam to the new steel and concrete state-of-the-art medical centre on the outskirts of that city. But the twentieth century dealt its harshest blow to the Musée d'Histoire Naturelle in the Jardin des Plantes in Paris. Here, man had finally put the finishing touches on God's creation. Until 1965, the Dutch writer Rudy Kousbroek could enjoy himself to his heart's content in this haven of natural history:

It was truly incredible what you could see in Le Musée: thousands, no, tens of thousands, no, hundreds of thousands, of stuffed animals, displayed in such a way that you immediately recognized the aesthetics of the eighteenth and nineteenth centuries: awesome; more inspired by principles of architecture and symmetry than by the systematics of science. You recognize it as that same spirit that infuses the first group photographs, of, for example, regiments in full regalia, bands, students at teacher training colleges, or members of influential families: faces to suit the occasion, enormous moustaches, the first

The Musée d'Histoire Naturelle in the Jardin des Plantes, Paris

row kneeling, the second sitting, the third standing, young men reclining to the left and right, heads supported on their hands, each holding the end of a scroll with an inscription, and flanked by indoor palms, Doric columns or Swiss Guards.

The way all of these stuffed animals are displayed evokes something else that originated during the eighteenth and nineteenth centuries: the full symphony orchestra, complete with brass, strings, percussions, choirs and soloists. The giraffes' necks are the double basses, the whales' ribs can be passed off as harps. Another feature which suits the collection very well is the deathly silence that envelops all the motionless animals, as if they are standing, waiting for a signal to bring them back to life. For the trumpets – not of the Resurrection, because that's already happened here, but of the Apocalypse.

Unfortunately, it was decided that this temple of resurrected nature had to be renovated, rearranged, adapted, restored to bits. As balm to the wound, though, there's still the Gallery of Comparative Anatomy further along in the Jardin des Plantes, which reminds you more of a skeleton tattoo than of an anatomical display. Surrounding this bony procession of thousands of skeletons are countless cases of anatomical specimens, averaging 150 years old and looking it. The true natural history museum is a museum of decline. There's no place for willows or birches; mahogany and walnut set the tone. And there's nothing that can stare better than glass eyes.

To see something like this in the Netherlands, you have to go to the remotest part of the country, to the eastern village of Denekamp, where, thank goodness, they've kept one room of the Natura Docet museum in its original state. It still smells of mothballs and formalin. On a sheet of glass, a stuffed baby crocodile has been crawling out of a stuffed egg since 1925. Animal heads hang everywhere. Many a specimen faithfully records the range at which it was shot: "Waterbuck between Amakoma and Beniaequator, 100 metres, July 1924, H. G. Lemon." Tacked, crookedly, onto the door post is a calligraphed note: "Elephant's tuft". The tuft itself has been missing since God knows

A stuffed cat in the Musée
d'Histoire Naturelle

when, but the adjacent cupboard contains a leg of the same animal. It's filled with colonial walking sticks. But for how much longer? Museums like the Natura Docet are becoming rare, as rare as the plants and animals they exhibit. Until the war, garden speedwell grew so prolifically around Denekamp that the director had no qualms about filling the museum's vases with it. Now the plant is rare and protected. Museums, however, are only rare; they seem too outdated to deserve protection. Modern museums have to be spacious, bright and open. Anything even remotely reminiscent of a museum is out of the question in modern museums. This is certainly the case in nature museums, whose subject is, after all, life itself. In such museums, it always seems to be spring. The old hand-blown jars and bleached specimens have long been taken home to the windowsills and linen cupboards of staff members and their friends. Their guests are amazed. In today's museums, the task of amazing visitors has made way for the challenge of educating them. Museums are now more like radio shops, with all their television screens and buttons. Fortunately, the switches are usually broken – as broken as all the stuffed animals. But let's keep that to ourselves. Birds, foxes and bats are stuffed as if they're alive and well. But why? You can see live animals in nature or, even better, on television. It's dead animals we need. They are what give nature its history. A lot happens between the moment a bird warbles on a branch and the moment it finds itself attached to it by a piece of wire. "Gift from the friends of the Museum. August 1928," says the label near the fossil of the long-stalked sea lily in Denekamp. Images loom up of men in outdoor get-up, women in muslin dresses on bicycles, garden spades in hands, watches on chains. The human touch is visible in each stuffed animal. Precisely because in the past they really did their best to make their work as true to nature as possible, the result is both nature and culture. It's not so much the way the bird once looked that you see, as the way people thought

Brown capuchin monkey (*Cebus apella*) with book, early nineteenth century

it looked. Birds have been elevated by collectors and taxidermists from anonymous creatures to specimens, and annexed into the human world. Animals that have been touched by real human hands are closer to us than all those crystal-clear television images, which are always separated from us by a television screen, a tele-lens and often an entire ocean.

A healthy natural history museum isn't ashamed of its roots: the curio collection. Nor should it be. During the seventeenth century, kings and princes, including the great Tsar of Russia, travelled to Amsterdam to admire the curio collections of wealthy merchants and learned scholars. The sole criterion for inclusion in these collections was extraordinariness. A curio collection is a glorified version of a boy's trouser pocket: everything unusual has been thrown into it. "Curtains of shame", the genitalia of exotic Hottentot women, lie next to dried sperm-whale penises; a rare crocodile hangs above an ancient Greek bronze. The most exciting items are Frederik Ruysch's anatomical specimens. The decapitated head of a child – topped with a lace

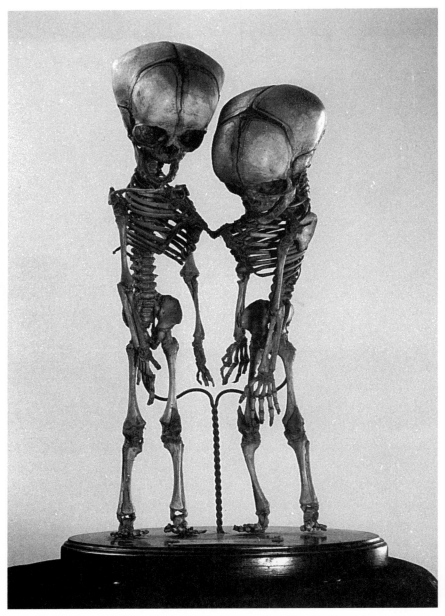

Skeleton of Siamese twins, prepared by Willem Vrolik

cap, its skin of an indescribably fine texture – looks out at the spectator from a glass jar full of secretly concocted formaldehyde, with the mysterious satisfaction of someone who, after a very short life, has continued to be around for a very long time. Equally amazing is the "specimen of the arm of a negro child with, in its hand, a specimen of the female genital". And, sure enough, a beautifully preserved hand, its nails seemingly just manicured, holds up a bedraggled pudenda for the viewers to see, as if it were an old-fashioned shop sign. In a curio collection, each new object amazes more than the last. Most people are familiar with this sensation from visiting the zoo. You never cease to be amazed by all the different animals. One has a very long neck, the other a nose that touches the ground, a third, antlers like the crown of a tree. But, on another occasion, in the very same zoo, you're struck by exactly the opposite: how much the animals resemble each other. They all have four legs, a nose out front, feet below. Similarities make systematization possible. It's a good thing, too, because otherwise life would be unbearable. Only by constantly organizing the chaos around him can man get a hold on the world and ward off madness. This is how taxonomical institutes emerged from

Frederik Ruysch's anatomical specimen of a child's head.

curio collections. Here, order and discipline reign; the rigid discipline of row after row of insects mounted neatly in boxes, of jars of appendixes next to jars of appendixes, all sliced into equal-sized pieces. Quietly mumbling in Latinate gibberish, man hopes to exorcize the chaos. But the displays don't satisfy the visitor: all those nine-banded armadillos next to all those nine-banded armadillos, those lesser spotted woodpeckers next to those lesser spotted woodpeckers. It's like a souvenir shop, with little windmills next to little windmills and wooden shoes next to wooden shoes.

A bad museum is like a sightseeing boat trip without a boat. A good museum is a place of pilgrimage. That's how it all began too. Long before museums existed, people were drawn to churches and cloisters. There, you could come into intimate contact with your favourite saint by kissing the ground he'd walked on, praying to his holy bones or giving money to the order he'd founded. In a good museum, you feel closer to the past, thanks to the traces it has left behind. I'll never forget the sensation I felt the first time I was allowed to hold a real dinosaur bone, a *touch bone*, made available to the public especially for that purpose. It was as if I were shaking hands with Adam and Eve. But I also remember the anger that possessed me when they confided in me that the bone was fake. Severe punishments should be meted out for exhibiting replicas in museums. To develop a relationship with the past, things shouldn't *look* real, they should *be* real. An ideal link was made in the 1980s, when the body of an extinct European bison that had lain in polar ice for 36,000 years, and then in a deep freezer for a few years, was thawed. It turned out to still smell distinctly of beef, "not unpleasantly mixed with the vague smell of the earth in which it was found and a whiff of mushroom". A dinner was organized and a short time later a dozen or so palaeontologists sat down together to enjoy steppe-bison stew. As a result, a dozen people are walking around today with living molecules in them of an animal species that became totally extinct 10,000 years ago. They've achieved the impossible: they've eaten time.

I have never had that kind of luck. But I have walked in the footsteps of a living dinosaur. A track was found in the Namibian desert of a species that had roamed those parts 175 million years before. That's an inconceivably long time ago. Sometimes, when I'm in a museum full of skeletons and models, I try to imagine such a half-eternity, but it never amounts to anything. In Namibia, I discovered what the problem was: those museum dinosaurs were already as dead as dodos by the time they fossilized, so that what you are looking at is a fossilized corpse. Those tracks that lay in front of me, on the other hand, were made by a dinosaur when it was in the prime of life. Identifying strongly with this, I jumped forward and placed my feet in its footprints.

The past not only lies behind us; it lies around us. Today's cockroaches look like yesterday's – from 200 million years ago, long before the first dinosaurs ever appeared on the face of the earth. They've refused to evolve.

Stubborn. The rise of the most advanced animal species – man – hasn't stopped them from continuing to live as they always did. Quite the opposite. It's precisely in modern metropolises where they feel most at home. They're among the liveliest of living fossils. That's why you scarcely notice how old they are.

The cockroach is a living fossil

Generally, the term "living fossil" is reserved for those plants and animals that have survived as the last species of a once plentiful group. Examples among the plants are the ginkgo and the *Welwitschia*; among the animals, the king crab and the *Nautilus*; and among the pubs, The Cheshire Cheese or The Last Drop. They all look so wonderfully prehistoric. King crabs look as if they've just crawled out of one of those schoolbook Palaeozoic landscapes. But what you don't see in their bizarre appearance is the fact that it's very common for an animal group to grow hundreds of millions of years old. Old forms simply continue to live alongside new forms. Past and future don't cancel each other out. Of course, some species do make way for new ones, but these by no means have to be the oldest. It's not a question of personnel management; old is not always bad. Just fall overboard mid-ocean and wait till a shark comes along and swims beside you. Sure, you could reflect on the primitiveness of such a fish – 100 million years old, no real bone in its bones, a ridiculous gastrointestinal tract and a completely outdated system of teeth replacement. Most likely, though, it wouldn't allow you enough time to finish those thoughts. Primitive or not, it's better adapted to the sea than you are. As long as the sea doesn't change radically – and it won't – there's no reason for the shark to change. And it's only a youngster compared with those organisms that live not only for, and around us, but also in us: intestinal flora.

"Intestinal flora" was one of the strangest terms I had to learn as a biology student. A flora in your intestines! My imagination ran wild. I felt the tulips

and geraniums in my stomach, as if they were the Dutch tulip fields themselves. Before long, drinking became associated with watering flowers. In
reality, of course, intestinal flora don't include tulips and geraniums. They
consist of bacteria, which, as invisible as they are infinite, perform their more
or less salutary work in the dark chambers of our body. But where does this
strange expression come from? It dates from the time when only two forms of
life were known: plants and animals. The two forms of biologists – botanists
and zoologists – also date from that period. Then, a third form of life was
discovered: bacteria and other microbes. Who was supposed to take care of
them? Is a bacterium a plant or an animal? And what about the amoeba? No
one knew. As if by the judgment of Solomon, the most mobile single-celled
organisms, with the thinnest cell walls, were assigned to the zoologists, while
the botanists were given the thick-walled homebodies: the bacteria. Since
then, we no longer have a zoo in our intestines but a botanical garden.

Much has happened in the intervening years. In *Full House*, palaeontologist
Stephen Jay Gould recalls how during his school days bacteria were totally
absent from the history of life. Posters of the oldest forms of life depicted

The Age of Reptiles. Nineteenth-century engraving

only ammonites and belemnites, the odd giant dragonfly and those strange ferns whose imprint, as a child, you sometimes found in the coal briquettes next to the furnace. The armoured forefathers of salmon and herring swam in water: that was the Age of Fishes. This was followed, in the next poster, of course, by the Age of Reptiles: a glut of dinosaurs, with the pteranodon, the first real bird, taking the place of the giant dragonfly. The climax was the third poster, the Age of Mammals, because it included us. We certainly learned that lesson well. But where did the bacteria go? During the first half of their existence – which lasted 2,000 million years – there were no fish, reptiles or mammals, only bacteria. So it would be fitting to call those first 2,000 million years, and, actually, the next 2,000 million years as well, the Age of Bacteria – because bacteria still exist, and in greater numbers and varieties than all the rest of life put together. Biologists are gradually beginning to realize this. The most important biologists today aren't zoologists or botanists but bacteriologists. During my youth, bacteria were only an appendix to a long list of plants and animals. Today, life is divided into 24 main groups, most of them full of bacteria. All of us animals are little more than a sideshow – one insignificant little phylum.

Thanks to their limited size, we can usually ignore bacteria and see them just as forage for toothbrushes and toilet cleaners, but as soon as we get diarrhoea and they gain control over our bodies, we're painfully reminded of the real balance of power. It's not a pretty thought to realize that such primitive little organisms are master over our highly developed systems. Nature is an uneasy bedfellow. Just watch people at the zoo, near the apes and monkeys: they laugh at their antics, but since we've known that we're descended from them it's no longer a very liberating laugh. We might have wished for a different family. We'd rather not be reminded of our origins – or of so many things.

For the best memories, it's better to have a bad memory. It's called nostalgia. One lick of Vaseline across the lens of your memory and your childhood looks idyllic, not to speak of your native country, the good old days, the time of the Raj. So it would seem to be a blessing that memory deteriorates with age. But that's an illusion. What declines – but not in everyone – is the

short-term memory. It only works as long as the brain reprocesses information: where you left your glasses, where you agreed to meet, what day it is
today. Or is it tomorrow already? The long-term memory seems to improve
the more time you have to poke around in it. It's not dependent on acoustic
stimuli; it's permanently engraved in your neurons. You'd be amazed about
the details from the past that can still be found in the remotest corners of
your brain. In search of vague phantoms that must be hidden away somewhere, razor-sharp images loom up of what's been lost forever. "Nothing is
worse", wrote Rudy Kousbroek, "than knowing your way around a house
that no longer exists." It's the opposite of a souvenir; it's an intangible
memory. The image in your mind is supposed to be a reflection of the world.
The more the two diverge, the closer you come to a mild form of madness.
Memories that have had their day should be replaced by other, more vivid
memories. Unfortunately, memories are as difficult to keep under control as
stomach aches or blushing. They show up when they feel like it. One moment
they bring a nostalgic smile to your lips, the next they sneak around in the
corridors of your mind like trespassers, until it creaks. But this doesn't make
it any easier to let go of what you once loved.

Animals don't look back, not even at the final farewell – death. Sheep
graze peacefully around a dead comrade, little dead birds are nudged over the
edge by their nestmates. Elephants are the exception. There are heartrending
descriptions of elephants that won't leave the side of their dead fellow
elephants, that root around in the dead mouth with their trunks, trying to
get the collapsed body back onto its feet, covering wounds with soil, not
taking the time to eat, swaying back and forth in misery for days on end.
You don't need a graveyard to mourn.

The veterinarian who was called to the side of a dead elephant in a safari
park in Florida a few years ago wasn't fully aware of elephants' obsession
with death. To establish the cause of death, he dissected the body on the
spot. He was quickly surrounded by mountains of intestines, towers of meat,
a landscape of entrails. Elephant was everywhere, all over the place, blocking
the path. To move the debris, a live crane was brought in: the bull with whom

An elephant mourns a dead companion. Chobe National Park, Botswana

the female had lived most of her life. First, he had to carry an amputated leg to a corner. He did this but was noticeably agitated. Then he was asked to roll away the head of his dead partner. He swayed back and forth, first on one leg, then the other, but did as he was told. He became so distressed from doing it that they finally let him out of the stall. The door was barely open when the bull stormed off as fast as he could to the remotest corner of the park, thrust his head onto the ground and trumpeted long and loud, until the keeper finally managed to console him.

People are like elephants. They don't accept the fact that their fellow man dies. But when the inevitable happens, the living, now called the surviving relatives, go in for remembrance. That's a lot of work, because with more than 100,000 dead people every year in the Netherlands alone, each with an intimate circle of roughly five friends, more than half a million people have their hands full with remembering. As clumsy as a herd of elephants, and without even a trunk to root around with, they seek consolation in each other's company, with coffee and cake and a shot of gin. It's a strange farewell that one witnesses at graveyards. We say the dead have departed, but the only ones who depart are us. It's more like an abandonment than a farewell. While the surviving relatives stoutly maintain they'll never forget the deceased, he's thrown away or burned like all the rest of our garbage. The Burial and Cremation Act doesn't allow anything else. That would be body desecration.

It's on account of the Resurrection that you're not allowed to desecrate bodies. You can't show up in bits and pieces at the Last Judgment. To amputate

A dog at a dog grave

a diseased leg from a living person is a great service to humanity, but to cut a dead leg off a dead person is blasphemy, to say nothing of skinning a person to stuff him. So even the most vile of criminals are made of wax in Madame Tussaud's. To see corpses made of real corpse you have to go to the mausoleums of communist leaders or the sarcophagi of Pharaohs. But the way their bodies have been conserved – like dead dodos, not true-to-life like the bluejays in the visitors' centres of the Forestry Commission – defies all rules of taxidermy. Only in London can you find a properly stuffed person. Philosopher Jeremy Bentham has been sitting there for more than 150 years in a mahogany case, hat on, cane in hand – at his own request, to avert the course of time, a souvenir of himself. He had already bequeathed his body to science at the age of 21, already chosen the glass eyes for his own human statue 20 years before he died. After his body had been publicly dissected, the remaining parts were reassembled and dressed again. If Bentham had had his way, it would be standard procedure today to have one's grandfather stuffed. Instead of bringing a picture down from the attic, you'd bring grandfather himself down to show your children. But that's a long way off yet. Bentham has seldom been moved from his permanent pew in the South Cloisters of University College, London.

Death masks are allowed, tawdry statues in parks, humiliating biographies.

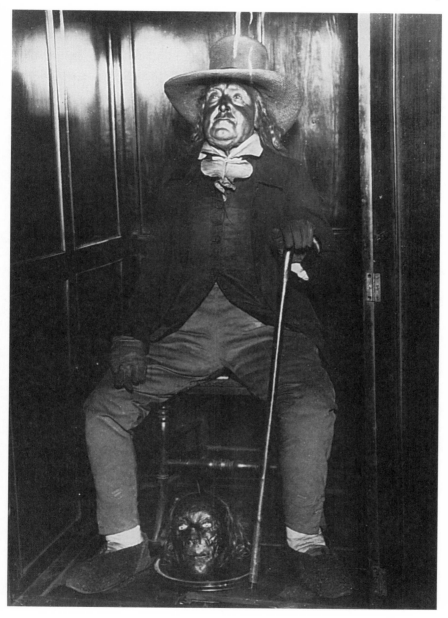

The stuffed philosopher Jeremy Bentham on display at University College, London

But skin as a memento is reserved for dogs. The Maison Deyrolle in Paris has devoted itself to stuffing pet dogs, cats, rabbits and parrots. This well-known firm, located between antique dealers and couturiers in the chic seventh arrondissement, offers us an alternative. For 300 pounds, you can have your pet immortalized in the position you remember him in best: ears cocked, obedient, being clever, sleeping, but never in the most obvious position – dead, belly up. That's not what remembrance is all about. Why don't we see stuffed dogs or cats on mantelpieces anymore, or next to armchairs, or in favourite baskets? A taxidermist who had already stuffed hundreds of dead dogs for their bereaved owners once gave me a possible answer. By the time a dog was stuffed, months later, its owners, their initial grief having long subsided, often didn't even come to pick it up. They already had a new dog. A live one.

The only way to get over the death of your dog is to get a new one. The only way to forget a war is to start a new one. Who still talks about the First World War now that there's been a Second World War? There still hasn't been a Third World War, so we keep mulling over the Second World War, as over a long-dead but – owing to circumstances – not-yet-replaced dog. There's nothing left to do but commemorate. And this we do: at monuments, in our thoughts, and, especially, on television. The battle of Stalingrad, the concentration camps, the bombing of Dresden, the last executions in Amsterdam by the Germans after the victory celebrations; I know them like the back of my hand. Despite all this, what continues to fascinate me are the eyewitnesses. You know, first you see the war hero in black-and-white performing his heroic deeds or the war criminal committing his heinous crimes. Then, suddenly, you see the same man or woman, this time in colour, 55 years later, sitting on a sofa in a flat in Moscow or Cologne. They survived. But what did they survive? The war? That isn't the most difficult thing as it turns out. With a bit of luck, you could escape from Adolf Hitler or Winston Churchill. Today, it's the Grim Reaper who is threatening – and he's much more implacable. As difficult as it is to detect five years of war on the face of eyewitnesses today, it is easy to see the next 50 years of eating, drinking, living and working. Time is worse than war.

Where have they gone: the war heroes, the superior Huns? Disappeared. Vanished. Literally gone up in smoke. What you see on television is only their present casing. In their bones, their heart, their kidneys, there's not a single molecule left from the cells they had in them 10, 20, 60 years ago. Nor is the blood in your veins the same blood you once had. You're not the same person you were in your childhood; at most, that's just a person you used to know well. Every part of your body has been replaced 100, 1,000, 10,000 times – by a replica. Broken cells or unrecyclable raw materials have simply been removed, exhaled, urinated out. The only thing that remains, the only thing that links you with your younger self and nourishes the illusion that all the photographs in your album are of one and the same person, is your memory. It's seated in your brain. And it is precisely that important organ that doesn't renew itself. While your whole body is constantly in a state of reconstruction, you drag your mind along with you like an attic full of increasingly dusty memories. So there's nothing left to do but to remember, remember and remember again. That is, to commemorate. Because what you are is none other than a memory of yourself.

9

EVERLASTING LIFE

It used to happen automatically. Today, there are even congresses and symposiums about it: building to last. All over Europe and America there are auditoriums and halls full of men with slightly larger hands than you'd expect to see at congresses and symposiums. Contractors and property developers are eager to learn. Building to last, that's synonymous with subsidies.

Building to last: God forbid it should ever happen. Imagine if all those houses from the 1960s had been built to last. Being saddled with rows of gallery flats and dismal housing estates until well into the twenty-first century. Let's first give serious thought to what they'll think of our buildings in 50 years' time before we dare to build anything that lasts. The same applies, *mutatis mutandis*, to 1,000-year empires, Superglue, triumphal arches and mayors-for-life. Transience is often preferable to permanence. Take plastics. Decades were spent trying to develop indestructible synthetic materials, then the environment got clogged, and today the hunt is still on for biodegradable plastics. This is how, if not the wheel, then at least the cycle of nature was reinvented. Without transience, it doesn't work.

Everlasting dictators, eternal love, perpetual sunshine: one is horror-struck. Everything with a beginning must have an end, otherwise there's no beginning. You don't need to wish your colleagues dead to see that are advantages to their having to step down one day. We prefer to reserve eternity for ourselves and our kind of people. The good among us. That's all right. Most cultures allow for the belief that life after death will be eternal bliss. Once, according to their myths, everyone was immortal. Holy books still testify to it. But then someone did something wrong – involving an apple or some such thing – and we were punished with mortality. But there's a remedy for that. A quick baptism, the occasional Hail Mary, polished shoes for Mass, a timely confession: these were the only things you had to do in my childhood to earn everlasting life. Not everyone received the same amount of mercy, I learned, but "God gives everyone at least enough that they'll be able to get into heaven". It was a great relief to hear that. But how to reconcile that with reality, where people don't live forever but die forever? You see mortal remains disintegrate into ashes and dust, so there must also be immortal remains that you don't see. Almost all religions have divided man into body and spirit. After death, the spirit goes off on new adventures of its own. It's an elegant way of ignoring death, which is still in vogue with millions of people around the world. It's not such a satisfying division though. Everything belongs in something else: without a glass you have no wine, without a body a spirit

Children in the Other World. Drawing by
Reginald Knowles, 1938

can't find its way around. Accordingly, the ancient Egyptians assumed that one day the spirit would return to the body. This meant the body had to be available. No everlasting body, no everlasting life. They could allow themselves to believe in this because of their climate. A body lasts for a long time in the dry sands of the desert, especially if it's mummified. To this end, Egyptian bodies were first pickled in natron for two months. Balsams were used to complete the chemical dehydration process. For safety's sake, the entrails were treated separately and interred in special urns. The Pharaohs lay more safely in the pyramids than they do today in museum showcases, with their sophisticated climate-control systems, which need constant monitoring. Museum visitors are amazed when they see bodies that have lasted for more than 5,000 years. What the display case doesn't tell you is that millions of Egyptians had themselves mummified in vain. When their spirits returned, they were greeted only by ashes and dust. Incidentally, the Egyptians were not the only ones – and certainly not the first – to mummify their dead. Near Arica, in Chile, mummies have been found that are many thousands of years older. It would seem that these corpses were first skinned and disembowelled, then dried over hot coals,

A mummified girl's head from Egypt

and eventually reassembled, if necessary being filled with minerals and clay. "A much more complicated method of conservation than that of the Egyptians", one pathologist describes it. But there, too, the desert did its bit to promote life after death.

Christians are less frugal with their bodies. In their book, the salvation of the soul comes first. The mortal remains are simply disposed of – albeit neatly, as becomes a dutiful citizen. After all, they're still God's creation. Sometimes He clearly lets this be known, too, as the following text, from *Gave lichamen* (Perfect Bodies) by Pater Timotheus, demonstrates:

The corpse of Saint Pascual Baylon, who died in 1592, emanated day and night a sweet-smelling perspiration, in such profusion that the attending physicians were faced with a great dilemma. Hundreds of people wiped the moisture off the corpse and spread it over the different body parts of the ailing, who were immediately cured. A blind man regained his sight after being treated in this way.

According to the Dutch bibliophile who dug up Pater Timotheus' obscure little work, "the absence of decomposition is the greatest mercy God can bestow on the mortal body; it is proof of the absence of sin, and for us, the living, a foretaste of the bliss of the hereafter." This wonderful fate was also reserved for Maria Magdalena of Pazzi, Catherine of Genoa, John of the Cross, Rosa of Viterbo and Christina of Holy Michael. The best-known perfect body is that of Bernadette Soubirous, who, with her Maria visions, laid the foundation for Lourdes as a place of pilgrimage. Her body is on display in a glass coffin in a cloister at Nevers. To save God embarrassment, He is helped by a layer of wax.

In the Netherlands, the Creator was assisted by "the Jesus of Eerbeek", who died in 1988 at the age of 78. He had stored no fewer than 250,000 stuffed animals in huts and bunkers all over the countryside. Like a second Noah, at God's behest he had collected samples of all species. As well as finding bears, deer, baby lions, black swans and crocodiles, the police, who discovered the bunkers in 1982, were especially struck by the masses of little birds, tissue-wrapped in cardboard boxes, like wise little old men and women from the Old Testament; ready for the Day of the Lord, the Resurrection. For Christendom, too, reunites the body and the spirit. Jesus Christ didn't require the building of pyramids or all that fiddling around with mummies. He would personally see to it that the bodies of the dead were reunited with their souls, and would rise up, alive, from their graves, as revealed by John the Evangelist:

> And I saw the dead, small and great, stand before God: and the books were opened; [. . .] And the sea gave up the dead which were in it; and death and hell delivered up the dead which were in them: and they were judged every man according to their works. [. . .] And whosoever was not found written in the book of life was cast into the lake of fire.

The bodies of the just will rise again in glory, together with the glorified body of Christ, but the bodies of the damned will rise in shame – with the terrible

Stuffed animals
belonging to "the
Jesus of Eerbeek"

stigma of eternal rejection. Because good or evil, reward or punishment, after the resurrection there is everlasting life: "[The body] is sown in corruption [and] raised in incorruption."

A God who offers you everlasting life: you'd be crazy not to believe in it, particularly as you get older and this temporary life becomes more temporary. The impression have under that churches are visited mostly by older people has been confirmed by Gallup polls in America. Two-thirds of the older people questioned found religion important in their lives and no less than 94 per cent of them were convinced God loved them, even though He wasn't always equally happy with them. Young people aren't so sure of this. The ageing of the church-going population certainly has to do with secularization, but that isn't the whole explanation. According to David Moberg, "Despite the decrease in attendance among the old-old that is associated with increased physical limitations and transportation problems, for at least half a century polls have persistently found the highest levels of church attendance to be among those past 60." Piety is a sign of ageing. Not only because older people have more time to contemplate life and death, but because of a growing need to make peace with their Creator before it's too late. If you take out insurance before going on holiday, it's certainly wise to get coverage for the journey to eternity. If for some reason that eternity doesn't materialize, your efforts won't have been wasted. People who believe in life after death are rewarded in the here and now with a better physical and spiritual condition. Which, of course, isn't to say that they're right. Even the most fervent believers aren't overly keen to depart this life.

Everyone wants to grow old, no one wants to be old. The trick is to grow old while staying young. But how? Polar-bear swims, gold elixirs, virgin blood, injections of finely ground dog testicles, holding your breath: everything's been tried, but the results haven't been spectacular. In ancient China, they sought the answer through sex. A school of Taoists laid the theoretical foundations for this at the beginning of the Christian calendar. According to them, the legendary Yellow Emperor became immortal after sleeping with 1,200 women. Normal people could live for 3,000 years, provided they adhered strictly to the rules. Preferably, sex should be had with virgins, and one wasn't enough: "A man has to have sex every night with three, nine or eleven women; the more the better." Elsewhere, people sought the fountain of eternal youth. You only had to bathe in it or drink from it to feel reborn. Shortly after Columbus discovered America, Juan Ponce de Léon thought he should seek it there. Instead of eternal youth, though, he discovered Florida, where to this day wealthy Americans are still looking for it. In Europe, we make do with a poor substitute, the health spa. People in search of health and youth come from far and wide to replenish themselves with the most trustworthy of waters. Even people who don't believe in anything drink Spa, Perrier and Buxton. There must be a tiny bit of magic in the stuff, otherwise surely they wouldn't dare to make it so expensive. Next to the bottled sources of eternal youth, "anti-death" food fills the shelves of health-food shops. What the Greek and Roman gods once achieved with ambrosia and nectar, surely we can match with bran and royal jelly? Unfortunately, life-prolonging remedies work no better than the love-inducing products from the sex shop on the other side of the street. And there's another similarity between source water and love potions: the fewer the effects, the better. After all, the only effects are side effects. Fortunately, most remedies have to be swallowed. The acid in the stomach knows how to deal with that.

As befits a reliable quack, miracle remedies are supported pseudo-scientifically. First, an explanation is given for ageing, then the remedy is palmed off on you. Every development in real science is followed to the letter. Once vitamins were discovered, miracle drugs were full of them. I still see the multi-vitamins they used to put on the corner of our dinner plates at every meal, summer or winter. We would grow from them; Grandma wouldn't die. No one really knew how these vitamins worked, but with so many of them in one little pill, the panacea seemed to be within reach. Today, with ageing being attributed to free radicals, vitamins are undergoing a revival. Vitamins A, C and E apparently clean up the free radicals. According to Nobel laureate Linus Pauling, we should consume a bottle of vitamin C every day. He swallowed twelve grams of it himself daily, 200 times the dosage officially recommended in the United States. He lived to be 92. No one knows what age he would have reached without vitamins, but we do know it's also possible to have too many vitamins and too few free radicals.

After vitamins, hormones became popular. Suddenly all problems could be injected out of existence. Old people have less of the sex hormone than young? Replenish it! Women in the menopause are given oestrogen; elderly men are pepped up with testosterone, as in the days of Brown-Séquard and Voronoff, although now without the accompanying monkey or cavia balls. If you lose hormones with your youth, they only need to be replenished for you to get your youth back. The same reasoning is applied to growth hormones. Old people have only half the amount they had when they were younger. No wonder they're old – because if anything resists decay, it's growth. Despite meagre results, one fad is hardly over and the next has already begun. After sex hormones, growth hormones and the wonder drug melatonin, DHEA (dehydroepianosterone) became all the rage. This adrenal gland hormone is converted into male and female hormones in the body. Apparently it's good for both the immune system and the vascular walls. Many people consider the DHEA-sulphate the long-awaited pill of youth. One person who vehemently disputes this is the discoverer of the substance, the French endocrinologist (and inventor of the abortion pill RU 486), Étienne-Émile Beaulieu. Unlike William Regelson, the American cancer researcher who has placed all his hopes in melatonin, he has his reservations about wonder drugs. "I'm 70", said Beaulieu in an interview, "and somewhere out there is someone, called Death, who has put a premium on my head and wants to see the end of me before too long. Just consider me an AIDS patient who, despite the lack of research and risks that new treatments bring with them, can't afford the luxury of waiting until science's jury comes in with a carefully considered verdict."

Health spas as the source of eternal youth

Once you're old, it's too late to grow old. He who wants to grow old has to start young. The recipe is well known: don't smoke, get more exercise and change your diet. Of these three things, it's the last that affects people most. With every bite you take, you let the wicked outside world into every last fibre of your body. It's logical: in the same way as the wrong fuel can make your car hit the dust, a bad diet is fatal to your health. The question is whether our diet is that bad. One thing we know for certain is that few lives have been prolonged because of radically altered eating habits. If macrobiotic and raw-food eaters really did grow a lot older than lovers of roast beef and Yorkshire pudding, their segment of the population would have increased accordingly. Even vegetarianism doesn't help. Extensive research in England has shown that radishes and parsley don't prolong life. There's no remedy on earth for decay, not even a herbal one. You should eat fewer, not more, carrots, together with fewer beans, fewer potatoes, fewer hamburgers, less ice cream and less cauliflower: you should simply eat less. Rats whose consumption is halved live twice as long, provided they get enough water, minerals and vitamins. It's all about calories. At the National Center for Toxicological Research in Arkansas, they managed to raise the oldest rat in the world by feeding it 60 per cent of its normal ration of calories. Methuselah the rat lived to be 55 months, three weeks and four days old. Elsewhere in America, in Baltimore, they've been conducting tests on apes since 1987. They, too, turn out to grow older and healthier on a calorie-restricted diet. A Dutch gerontologist is less enthusiastic about the results in people. According to him, human lives don't really last longer because of fewer calories. They just seem to.

Is more exercise better than less food? Is it possible to train yourself to live forever? This is a difficult question to answer because sport has been around for such a short time. Animals don't train and your great-grandfather had something better to do with his time. Sport is one of man's more recent creations.

Man has never worked himself up into more of a sweat than since he invented sport. Sisyphus turns physical! All the work that was taken off his hands by machines is nothing compared with all the training, jogging and competing that's going on in stadiums and sports centres. Fitness centres are filled with equipment for training muscles to lift weights, turn wheels

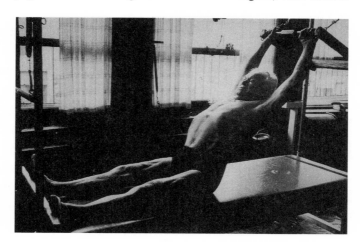

Is it possible to
train yourself
to live forever?

and work pulleys, as in the days before sport was created. But there's one difference: they serve no purpose. A spanking new industrial site could easily be hooked up to all the energy produced by all the fitness centres in England. Conversely, a sawmill could be converted into a fitness centre just by pulling out the plug. But fitness freaks aren't interested in that. They're busy stretching their bodies and their lives.

Getting old is no fun. Eating less and giving up smoking is torture for a lot of people. Physical labour is torture too. It doesn't take much imagination to see a fitness centre – with all its stretching equipment – as a torture chamber. Torture is timeless. We don't need to wait until fitness centres celebrate their 100th anniversary to know if torture prolongs life: in 1977, Statistics Netherlands presented a study on the survivors of the German concentration camps. To determine the effects of mistreatment, the researchers looked into how quickly the survivors died after the war. The results were astonishing: after being liberated, the lives of the camp survivors were not shorter but longer than other Dutch people's. Among the younger survivors, each year a third fewer died than was expected; among the older survivors, it was half as many. Whoever had managed to survive inside the camps had little difficulty staying alive outside them. This can be seen as a sinister case of natural selection. Only the tough in body and strong in spirit had managed to survive. Another of the researchers' explanations was that after the war, the survivors – embittered or not – had withdrawn from the welfare society. It is generally accepted that one lives longer by living frugally. It's not just by chance that hermits, stylites and ascetics are invariably depicted as old men. If one gets sick, weak and nauseous from everything that's delicious, one only has to renounce everything that's delicious to live, if not long and happily ever after, then at least long. The sixteenth-century Venetian Luigi Cornaro lived both long and happily from this. When, at the age of 50, he realized the effect his dissolute lifestyle, insatiable greed and licentious behaviour had had on his body, he switched to a Spartan existence from one day to the next. He reached the age of 98 – on a frugal diet of bread, meat, bouillon and wine. "The food a person resists after a solid meal does him more good than the food he has already consumed," he wrote, in his *Discorsi della vita sobria* (*Discourses on a Sober and Temperate Life*). He enjoyed his old age, if for no other reason than that his book was a great success. After he died, that success became even greater. During the eighteenth and nineteenth centuries, in England alone it was reprinted 50 times, and it can still be bought in America. That won't surprise people with any insight into human nature. We are, after all, mad about all the things we don't love. History is full of self-chastisement, voluntary fasting, Spartan lives and self-punishment. Where there's crime, there's punishment; he who feels criminal quickly becomes his own hangman. Although malnutrition mostly affects the southern hemisphere, it's in the northern where most people are hungry. Surrounded by restaurants and fast-food outlets, modern northern man is forever on a diet. Not to lose weight, but because he feels guilty about being fat. You don't lose weight from dieting. That's been proved repeatedly. This is because you get hungry from dieting, and when you're hungry you eat. And from eating you don't lose weight, you gain it. You feel guilty about gaining weight and because of that guilt, you go on a diet. You just can't win!

The older you get, the guiltier you feel about everything you've done in your ever-lengthening life. For fear of decay, many people spoil their last years with diets, fitness exercises and the fear of God. Is it a long life you want or a fun life? Or an acceptable compromise? Do the costs outweigh the benefits? It's a question of quantification. Stopping smoking yields the highest benefits with the lowest costs. Every minute that you no longer smoke is added to your life like a bonus. Even the tobacco industry understands that you don't grow old from smoking; all you ever see in their advertisements are young people. Eating healthily prolongs life less than not smoking, but a few additional years never did any harm. The question is whether the benefits still outweigh the costs. Stopping smoking is a torture that lasts for weeks or months; healthy eating is, for most people, like a life sentence. If living longer means you're miserable for longer, it's time for meat pies and cream cakes. Fitness exercises can't save you anymore. If muscle power gave you longevity, homes for the elderly would be filled with bricklayers, boxers, lumberjacks and professional footballers. But they aren't. Working your body into the ground doesn't stave off ageing. It does help against heart and vascular diseases though. In R. S. Paffenbarger's study, ex-Harvard students turned out to live two years longer if they'd been active in sports. But a short jog around the park wasn't enough; a dog-trot is for lame dogs. To reach 82 instead of 80, the Harvard graduates needed to have worked off 2,000 calories every day for a period of 60 years, the equivalent of heavy dock work for a full two years of one's life. The 100 weeks their lives had gained were spent huffing, puffing and sweating. If you enjoy scaring the living daylights out of respectable amblers by wearing a candy-striped jogging suit, by all means do so. For all the other people who want to live long, what counts is: act normally and hope for a bit of luck.

The fact that we live twice as long as we did a hundred years ago isn't thanks to doctors but to plumbers. Most disease-causing germs were eradicated with the monkey wrench. Doctors helped too, with their injections against infectious illnesses. But no plumbing or penicillin helps against cancer or heart attacks. Heavier artillery is needed to combat these. One day we'll probably have it, too, and it will mean a few extra years of life. But if we go on as we are and eradicate all illnesses, does this mean we'll have everlasting life? No, of course not. If you eradicate killer number 1, then killer number 2 becomes killer number 1, and so on, until you stumble on the illness called life. A centenarian hasn't managed to last because he's escaped every illness. No. If he's opened up, you'll see he's got every disease in the book – but not one

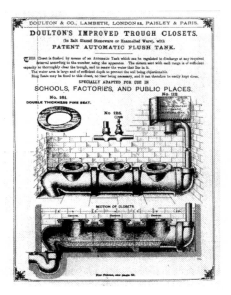

The fact that we're living longer and longer isn't due to the doctor but to the plumber

that dominates. Each part of his body is as worn out as the next. Everything goes smoothly for 100 years and then suddenly, in a chain reaction, everything breaks down. But it would make quite a difference if there were no clearcut fatal diseases anymore. If we removed the 20 most common causes of death – from cancer to murder – we'd gain 20 years, and centenary celebrations would be commonplace. But the costs would be astronomic. The reason it's been possible to wipe out infectious diseases is because vaccinations and anti-biotics are so cheap; heart transplants and artificial kidneys fall into quite a different price category. In Western Europe, the cost of healthcare has more or less doubled since 1980, but life expectancy has nowhere near doubled. Every month gained costs the healthcare system a fortune. It's money better spent elsewhere: on education, for example. That way two birds can be killed with one stone. Young people grow smarter and old people older. A good education increases life expectancy by three to four years. This effect becomes much greater if you look at it in terms of healthy years. Better-educated men enjoy ten more years of good health than poorly educated men, who have to suffer no less than 21 years of ill health before they're finally allowed to die. But good education, good plumbing, good food and good healthcare only exist thanks to the affluence of our society. So we're caught in a vicious circle: from affluence you get the diseases of affluence and the diseases of affluence shorten your life. The circle will only be broken when there's a fitness centre next to every industrial complex and a food advice bureau next to every restaurant. But that requires even more affluence, and so the circle is complete. If everlasting life does exist, its costs are prohibitive. Both for the State and for its citizens.

A citizen pays the highest price for everlasting life: death. He only lives on if he gets cancer. At least, that's what happened to Henrietta Lacks in the early 1950s. A few weeks before she fell ill, this American woman gave birth to her fifth child, a son. A few months later, on 4 October, 1951, she was dead. But her cells still exist today in laboratories all over the world – under the code name HeLa. Today, 400 times as many of Henrietta's cells are in circulation as she ever had herself. And these cells have experienced more in life than she – as a simple black woman – could ever have hoped to. They've contributed to polio research, been into space along with two white mice, and, through their tissue culture, saved the lives of hordes of mice undergoing allergy tests in cosmetics laboratories. That her cells have become so much more famous than their owner has to do with their disease. It began as cervical cancer, but when Henrietta lay on the dissection table, the assistant saw "tumours in all her bodily tissues, wherever you looked". Within six weeks, the cells divided outside the body at the record speed of once every 20 hours. Everlasting life is in a big hurry.

Henrietta Lacks whose cells live on under the code name HeLa

Temporary life isn't. Otherwise it would become even more temporary. If an embryo were to grow at the same rate as a cancerous tumour, every pregnancy would risk ending in an explosion. It would be over very quickly, not only for the mother but also for the child, because each cell has only a limited number of divisions. After 40 or 50 divisions, Leonard Hayflick discovered in 1965, the cell stops dividing. At the rate of the HeLa-cells, the embryo would be a wrinkled old specimen within six weeks of fertilization; old and weary of life, with nothing left to divide. Normally, cells last a lifetime, and that's not coincidental: including all the cells one sheds and loses in the course of one's lifetime, 40 divisions per cell are enough to produce an adult body. After that, the process starts to fail here and there, until vital bodily functions are brought to a halt and it becomes a hopeless case. But if cancer cells can divide *ad infinitum*, why can't healthy cells? Why does the body let its healthy cells die? Probably for fear of cancer. The unchecked growth of cancer cells instils so much fear in the body that a system of checks has been introduced into each cell. In cancer cells, one gene is turned on that is turned off in normal cells. Because of this, normal cells don't make telomerase. This is an enzyme that inhibits the fraying of chromosomes. Without telomerase, the ends of the chromosomes (the telomeres) fray progressively after each cell division until eventually there's nothing left. The cell can't distinguish properly between the worn ends and the broken chromosomes, and so it does what it always does with broken chromosomes: it glues. With the chromosomes glued together at the frayed ends, the nucleus can't do its work and the cell packs it in. In theory, it should be possible to make normal cells as immortal as cancer cells. You only have to turn on the turned-off gene. The switch can be found too. That's not the problem. The problem is how to turn it off again before the cell outperforms itself and divides until it bursts. Should we dare do what the body itself does not? Perhaps not, but once the switch is within reach, we probably won't be able to keep our hands off it.

Yet death is not invincible. Even if you don't believe in Jesus Christ and Lazarus, there are those who have defied it. They're probably sitting on the windowsills of your house right now being immortal. Geraniums, begonias and spider plants (Chlorophytum) have no fear of the grave. They allow themselves to be propagated by cuttings. Every time this is done with a plant, death is defied. A whole new being grows out of the cutting, which can be cut again, and so forth. The mother plant never dies. It lives on through the cutting and again later, through that cutting's cutting. In every cell of every generation, it's still the plant it was. Here, birth and death lose their meaning. It's true that many plants become weaker after repeated cutting, until finally you have to buy a new plant at the garden centre. But that's not the case with spider plants. They cannot deteriorate. Each of the little spider plants that sprouts out of the effusive clump of mother grass turns into an impeccable daughter plant. There's no money in this plant for garden centres, and it doesn't really need us either. In the wild, the spider plant propagates itself when the satellites touch the ground and take root around the mother plant like a complete solar system. Strawberries grow this way too. They know how to do it – reproduce using flowers and fruit – but when they're in a hurry, they play conquer-the-countryside with their runners. Field bindweeds do it underground, with rhizomes, which give the plants new living space and de-weeders headaches. Creosote bushes spread completely asexually. In the

Mojave Desert in California, one specimen has been doing it for more than 10,000 years. A better known example is the Cabernet Sauvignon. This grape has been propagated continuously from cuttings for more than 800 years. And bananas are so prudish they haven't produced fertile seeds for hundreds, if not thousands, of years. They prefer to let us propagate them into eternity using cuttings.

Some animals can be propagated by cuttings too, but then it's called budding or binary fission. Jellyfish lift off the top of the pile like little saucers; polyps begin as warts on their mother's skin before assuming hideous shapes and separating to set themselves up independently. These animals are just as immortal as the spider plant. What is so enviable is that they don't need complex technology or highly developed brains to achieve this. On the contrary, the lower the organism, the greater the chances it will live forever. Blessed are those with primitive body structures. Among the highest organisms, only a single salamander can reproduce itself asexually, but among the single-celled organisms, it's the order of the day. Paramecia and flatworms divide like bacteria. Because the two daughters are identical, one isn't older than the other, and the mother cell isn't the mother in the sense of being older. Being as young as your daughter is an ideal that was achieved long before the first person – let alone the first beautician – ever appeared on earth. Of course a bacterium can die, and for it, too, death has many faces – being eaten by a blood cell, brushed to death by a toothbrush, or drying up, water droplet and all – but dying from old age isn't one of them. Every bacterium originates directly from the fission of the Eve of all bacteria. All living bacteria are thousands of millions of years old.

In fact, the cloning that makes biologists so proud and people so frightened is none other than propagating through cuttings. What's wrong with it, aside from all the fuss and bother? The pitfalls became evident with the cloning of Dolly, the Scottish sheep born – or whatever it's called for cloned animals – in 1996. After only a year, her cells showed all the symptoms of ageing. This isn't so strange because she was cloned from the genetic material of an adult animal. You can't clone lamb from sheep. That's not what consumers are waiting for. They want lamb. So how do you make lamb from sheep? This is something every butcher would like to know. For the answer, ask any farmer: you need a ram. Put a ram with your sheep, and within the blink of an eye you'll have more lamb than you'll know what to do with – neatly bundled in wool, frisky legged – a field full of them every spring. Reproduction is the ideal rejuvenation treatment. That's obvious – but we act as if it's the strangest thing we've ever heard of. In Japan, they understand it better. There, they haven't restored the Shinto temple in Ise; they've rebuilt it – every 20 years since the fifth century.

Dolly, the cloned sheep

If they hadn't, repair work or not, it wouldn't still be with us.

To become young successfully, reproduction has to be sexual. China's Yellow Emperor didn't have to be told this. Unfortunately, it isn't you who becomes younger but your offspring. Two old people together make a young child. It's like a miracle, equal to that of the everlasting life of a Christian after death. The solution is similar too. You have to split the person. In the same way as believers think they are made of mortal remains and an immortal spirit, August Weismann made a distinction in 1885 between the mortal soma and the immortal germ line. Normal body cells (the soma) are just as transient in Catholics as they are in Protestants. It's only the germ line – which makes its way, as an egg cell or sperm cell, from one generation to the next – that's everlasting. From the point of view of the germ cell, the body is no more than an instrument for making new germ cells. The germ plasma, which produces the egg cells and the sperm cells, is as immortal as the Holy Ghost. As if to emphasize the body's insignificance, the germ plasma is kept isolated in the embryo from a very early stage – from the very first week in mammals – and disposed of as soon as it has produced the next generation. The old specimens have to clear the field as soon as the young specimens appear. After all, young ones can reproduce better than their fathers and mothers, who increasingly get underfoot. Some researchers believe that each cell has its own timing mechanism, which induces it to self-destruct at the right moment. Such a thought is romantic but unnecessary. If cells are poorly enough constructed, they'll clear themselves away. Surely manufacturers don't put expensive time bombs in their cheap watches to improve sales? The trick for the manufacturer is to make sure that the watch stops working after the warranty has expired, preferably as soon as possible after that. Aside from those who die of a heart attack or a brain haemorrhage, most of us don't die as if an alarm went off. It's not all over the moment your last ovulation takes place or your last child leaves home. After the warranty has expired, there's a welcome reprieve, built in by the Great Watchmaker as a precaution against expensive repair claims. Eventually you'll stop working anyway. Meanwhile, new watches that have been fully wound wait to tick out their days. Whoever has a new system for staying young doesn't need an old system anymore. So get rid of it! To cite Alfred Russel Wallace, discoverer, together with Charles Darwin, of the theory of evolution by natural selection: "Thus we have the origin of old age, decay, and death; for it is evident that when one or more individuals have provided a sufficient number of successors they themselves, as consumers of nourishment in a constantly increasing degree, are an injury to those successors. Natural selection therefore weeds them out, and in many cases favours such races as die almost immediately after they have left successors." Death is the price we have to pay for sex. A high price, no doubt, but at least you have something to show for it.

The idea that germ cells are worth more than the mortal casings they produce is also gaining ground in zoos. Why bother maintaining expensive zoos, full of animals that defecate, keepers that ask for pay rises and whining children, when you can also store species in the form of everlasting germ cells? At least four zoos in the world have switched to the deep-freeze. Egg cells, sperm cells and – currently, for the sake of convenience – very young embryos are lying in these frozen zoos waiting until they're needed. Instead of the male panda, today male panda's sperm is sent out into the world – at 196 degrees Celsius below zero. In the future, even that may not be necessary. To keep a species alive, it's sufficient that germ cells exist. The rest is circus, sentiment and sensation.

If body cells wear out like the cogwheels of a watch, how do germ cells manage to stay young forever? It's partly thanks to telomerase. As in cancer cells, the gene for telomerase is turned on in egg cells and sperm cells. This guarantees everlasting life. But what about their youth? Like so much in nature, this is influenced by natural selection. In the same way as forests are the arena in which the best animals will prove themselves, the body is a jungle where only the best egg and sperm cells will achieve their goal. Of the million egg cells that a woman is born with, only a few hundred are given the chance to develop. Only the best ripen, settle in the uterus and unite with a sperm cell. And even then, many fertilized egg cells abort prematurely because of small defects. In humans, 100 million sperm cells are produced every day. To this end a man crochets enough DNA every 24 hours to encircle the earth twice. To limit this never-ending potential for error, all sperm cells undergo trials that make the quest of the Knights of the Round Table pale in comparison: only the very bravest can ever enclose an egg cell with its flagellating tail. And even that isn't enough for nature. Since errors in germ lines have much more far-reaching consequences than errors in body cells, germ cells are checked and repaired much more rigorously during the production process. It's an assembly line of the kind any consumer organization would wish for: firstly everything possible is done to avoid errors, and subsequently the products are subjected to such mercilessly strict tests that any weakling will fall by the wayside.

So you would think the best thing nature could do if sex is involved would be to allow as many young specimens to be born as quickly as possible and to dispose of everything that's old. If this was the case, nature would feel as uniform as everything that was "Made in Hong Kong" in the 1960s. In nature, however, a principle reigns that many managers refuse to understand: the

X principle of the alternative strategy. What this boils down to is that something can be done this way but also that way. This and that are often contrary yet equally good. Children's questions make this clear.

"Daddy, why is an elephant so big?"

"An elephant is so big, my son, because then he doesn't need to be afraid of anyone. That's why there are elephants."

"Why are mice so little then, Daddy?"

"Mice are so little, my son, because that way they can always dive away into holes. That way they don't have to be afraid of anyone. That's why there are mice."

Daddy's right. It doesn't matter whether you're a big elephant or a little mouse. Both strategies have their own advantages. The only thing you mustn't be is a little elephant or a big mouse. You have to make a clear choice: subtly camouflaged or strikingly coloured, super-conservative or ultra-modern, immediate escape or instantaneous attack. The principle of alternative strategies has been studied best in reproduction. We have our own way of doing

A stork family

it; we produce a limited number of children and take such good care of them that there's enough of them left to take our place when we die. Birds do it this way too. That's why some television networks are so eager to broadcast nature programmes about bird's nests – Daddy bird and Mummy bird take good care of their little baby birds – it's an ode to family life. But birds and people are exceptions. Most animals do precisely the opposite. In a most un-Christian fashion, they produce as many offspring as they can with as many partners as possible and then take off as fast as they can. An extreme example are cod. They discharge their eggs and sperm randomly into the water. The germ cells have to find each other all on their own. To give them at least a fighting chance, the females produce hundreds of thousands of eggs, and the males discharge possibly an even greater number of sperm cells. Even if only one of them reaches its target, the groundwork has still been laid for offspring. Herring do it this way too, together with sea cucumbers, coral polyps and blennies.

We're not fish. We take care of our handful of offspring for a long time. And that takes time. Add to that the time we need to grow up and you'll understand why we're among the longest-living species. Contrary to the complaint often heard among the elderly, we don't place all our bets on youth. It takes so much effort for people to be assembled, to grow up and to learn everything they'll need for later that it would be a shame to discard them again right away. You'll always find plenty of supermarket shelf-stackers; physics professors are harder to come by. It's a question of investment and returns. When a person is finally finished, one of two things can be done: invest in repairing what you already have or invest in producing new, flawless children. Old age always loses in the end. British anatomist Peter Medawar knew why as far back as 1952. There are always more young people than old. It can't be otherwise, because old people have been exposed to dangers over a longer period of time; it doesn't matter whether it was in the form of lions or cars or ultraviolet radiation. In a world where there are more young people than old, the needs of the young dominate. Features that were advantageous when you were young continue to exist, even if they become a source of trouble when you're old. Old people reproduce so seldom that natural selection no longer has a hold on them. Youth always wins. How wonderful that we were all young once.

People are very good at prolonging life, like children who refuse to go to bed. As a child, finding myself in bed despite this, I used to dream of a world where the day never ended and you could stay up forever playing. Adults have frequently recorded such childhood dreams, but none so tellingly as Jonathan Swift. During one of his journeys, Swift's hero Gulliver hears of people who never die, the Struldbruggs. Gulliver wouldn't mind this for himself:

> [...] if it had been my good fortune to come into the world a Struldbrugg ... I would first resolve [...] to procure myself riches [...] and to apply myself to the study of arts and sciences. [...] Add to all this, the pleasure of seeing the various revolutions of states and empires, the changes in the lower and upper world, ancient cities in ruins, and obscure villages become the seats of kings. Famous rivers lessening into shallow brooks, the ocean leaving one coast dry, and overwhelming another: the discovery of many countries yet unknown.

Gulliver's hosts can hardly suppress their laughter on hearing such naïveté. Yes, they know that people everywhere else want to live for a long time, "that whoever had one foot in the grave, was sure to hold back the other as strongly as he could. That the oldest had still hopes of living one day longer", but they know better here. They know the truth about the Struldbruggs:

> [Struldbruggs] commonly acted like mortals, till about thirty years old, after which by degrees they grew melancholy and dejected. [...] When they came to fourscore years [...] they had not only all the follies and infirmities of other old men, but many more which arose from the dreadful prospect of never dying. They were not only opinionative, peevish, covetous, morose, vain, talkative, but uncapable of friendship, and dead to all natural affection, which never descended below their grandchildren [...] and whenever they see a funeral, they lament and repine that others are gone to an harbour of rest, to which they themselves never can hope to arrive.

The immortals are right to be jealous of the mortals, because:

> ... at ninety they lose their teeth and hair, they have at that age no distinction of taste, but eat and drink whatever they can get, without

relish or appetite. The diseases they were subject to, still continue without increasing or diminishing. In talking they forget the common appellation of things, and the names of persons, even of those who are their nearest friends and relations [. . .] they can never amuse themselves with reading, because their memory will not serve to carry them from the beginning of a sentence to the end . . .

Gulliver thinks again. He's ashamed of his naïve fantasies and understands that "no tyrant could invent a death into which I would not run with pleasure from such a life". He thought it a good idea to take a few Struldbruggs back home with him "to arm our people against the fear of death".

The story about the Struldbruggs draws heavily on the Greek legend of Tithonus, a Trojan in love with Eos, the goddess of dawn. Eos asked Zeus to make Tithonus immortal, but forgot to ask that he be granted everlasting youth at the same time. Fortune smiled upon them until Tithonus grew old and, in the end, withered away in a little room where he was destined to wither away forever. All these centuries later, neither Gulliver nor Tithonus' warning has lost its significance. Everlasting life is like paradise or the land of milk and honey: something to long for and dream of but not really visit. Anyone who's ever been to Butlin's Holiday Camp knows what I'm talking about.

Jonathan Swift has been dead since 1745, but he lives on in his stories. Unfortunately, we can't make a new Swift. His books can still roll from the press, though, at a rate of thousands per hour if necessary. So religion turned out to be right after all: the body dies, but the spirit lives on – as a message for those who stay behind:

> father once bought
> a collected work:
> a selection of poems
> of striking quality.

Arthur Rackam's depiction of the Struldbruggs, who lived forever,
in *Gulliver's Travels* by Jonathan Swift

next to what he liked
he put a dash
the odd time
an exclamation mark

occasionally
I re-read that
very concise
biography:

a code
of dots and dashes
reveals where
he found delight.

WILLEM WILMINK

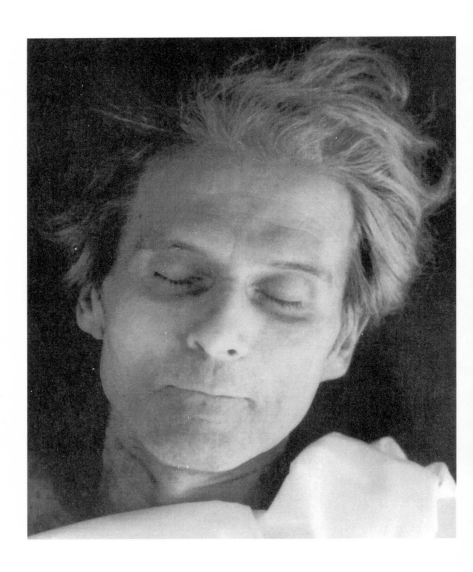

IO

DECAY OR FULFILMENT?

"It is finished!" And, bowing his head, he gave up the ghost, so that the Scriptures could be fulfilled.

This is how the Bible ends. The whole of the Flood, all of Egypt's ten plagues, the sacrifice of one's own sons, all that begetting of this person or that, Sodom and Gomorrah, the miraculous feeding of the 5,000 – they all reached their climax on the Calvary. After hundreds of pages of patriarchs, whoremongers and Pharisees, the sins of the world are gone, the prophecies fulfilled, mankind saved. As an epilogue, Jesus rises again. *Deus ex machina*. Happy ending. With a sigh of relief, you close the book.

All of Christianity revolves around this end, which is celebrated – this year for almost the 2,000th time – every year anew. Masses are held, the Pope gives thanks for the flowers, shopping malls do good business. The Lord has truly arisen. But Easter can't hold a candle to Christmas. While, for Christmas, cities are decorated and wars stopped, Easter, with its meagre chocolate eggs, is over before you know it. You don't receive any expensive presents, trees are left in peace in the garden and no one – fortunately even in America – has ever heard of Easter songs. We would rather celebrate Christ's birth than His death, even though that's what it's all about. A promise is more fun than fulfilment. Our hearts beat in happy anticipation. Goals are there to be looked forward to. It's no coincidence that Christmas falls on 25 December, when the days start to lengthen again, as a preliminary announcement that a new year is about to be born.

It's the same old song every year: spring breaks out. Little lambs frolic in the meadows, weeds send up their cotyledons like little periscopes, mosquitoes prime their piercing mouth-parts, buds burst like infected wounds. The whole of nature is rebuilt again from scratch, like a marquee. For her umpteen-millionth spring, Mother Nature turns on the tree root pumps, hedgehogs mindlessly cross roads, birds clear their throats, nests are built, young people get ideas. While the neighbour lets his new lawn-mower purr, the abattoirs start up their spring-lamb slaughtering machines. Optimistically dressed sun-worshippers catch nasty colds. All of nature whistles, peeps and hums. New life spreads, like an epidemic. Nature awakens.

The question is whether all this is fun. Who likes getting up? A normal person prefers to roll over. Mother Nature is normal. She sets to work again with nothing less than reluctance. And for whom? Owls are happy to see the mice coming, but the mice would rather see the owls going. The owls are hungry, the mice are hungry, all of nature thirsts for water, blood and knowledge. Everything that's alive craves and indulges. In spring an incredible disquiet takes hold of the earth.

But this doesn't happen voluntarily. Why should you build a nest if you're a bird? Why should you let yourself be eaten out of house and home by your own brood? Why should you risk your life for theirs? If birds had any say in the matter, they'd refuse the job. But it's not only up to them: hormones have taken hold. As if once in a lifetime weren't enough, birds have a new adolescence every spring. That's why they show off so much, like drunk teenagers outside pubs. Stunt-flying, they whistle to the females. Something's about to happen. They can feel it. But what is it? Not a feather on their heads suspects that in a few weeks' time they'll be slaving away like dutiful fathers in the name of offspring of their own making. But first there's a nest to be built. Not because – like loving couples at Mothercare – they're looking forward to a cosy clutch of youngsters, but because they're too dumb to build something that will last for several years. There's no time to waste either. Just back from Africa and the hormones start acting up. No such thing as catching one's breath. To work! Now! Damn. What we call spring is nothing other than nature waking up in the morning in a bad mood.

People enjoy walking in nature's bad mood. In the Netherlands in spring, people take off by the busload to the Betuwe, the traditional fruit region along the Rhine, in search of the orchards they once saw blossoming there. Guided by apple, cherry and pear route signs, they manage to gather enough glimpses of the blossoms for a spring feeling to come over them. When it rains, the touring buses still show up. They were reserved in advance. You'd be amazed how many pensioners can fit into one bus. And just watch them enjoy all that beauty, especially that anticipation. Every blossom holds a promise. One day, all those blossoms will be apples, cherries or pears. That's what it's all about. It's because of pears that pear trees have blossoms and pear farmers subsidies. It's an added advantage that tourists come to see the blossoms. Growing crops and milking the tourists that come to admire them: it's a nice form of mixed farming. But tourists alone aren't enough to keep the pot boiling. They only come in spring.

You don't see a single tourist in the Betuwe in autumn. By the time the apples and pears have reached their prime, there's no one left to enjoy their beauty. Unappreciated, they dangle from their branches until plucked by indifferent hands and removed to dark cold sheds. Even though it's still summer when the harvest begins, none of the fruits of the farmers' toil are of any use to the Tourist Board. Apple and pear routes aren't meant to take you past rows of apples and pears. You see those every day at the greengrocer's. People only want to enjoy the early days of the fruits-to-be. It gives them the same feeling as little lambs cavorting in meadows or babies' moon faces peering out from under the blankets of a pram.

If autumn has one distinct feature, it's that it persists. In temperate zones, it rains and storms, strong gusts of wind cause alarm, people become depressed. One out of every ten people feels tired in autumn, is constantly hungry and can't be dragged out of bed. More alcohol is drunk than is wise. In northern countries like Sweden, where it always seems to be autumn, alcohol abuse is rampant. This has nothing to do with the lower temperatures, because in Hawaii, where it's always pleasantly warm, people also get depressed during this season. It seems to have something to do with light. Many patients feel much better after half-an-hour under a bright light. It doesn't matter whether that half-hour is in the middle of the day or at some other time. What seems to count isn't the length of the day but the intensity. Naturally, there are those who enjoy autumn, but they're generally the more contemplative types, who consider "loving spring" child's play, something for beginners, comparable to "loving dogs". It's easy to love something that loves you, but as seasoned cat-lovers know, nothing beats love which – totally randomly – may or not be reciprocated. I wouldn't be surprised if studies showed that autumn-lovers were to be found mostly among cat-lovers.

The biggest autumn-lover is nature itself. For nature, autumn is the king of all seasons. There's no other time of year when so many goals are achieved, so many expectations met. At last, the trees and shrubs are filled with berries and nuts. The complex time-consuming process of budding, germinating, pollinating and fertilizing is finally over.

By the time the apples and pears have reached their prime, there's no one left to enjoy their beauty

Relieved, trees give up their now-superfluous leaves to the wind. Seed-loving animals imagine themselves in the land of plenty. From far and wide, finches come to our pine forests to peck the cones empty. Other birds, having hatched and raised their young, feel they deserve a holiday and head south. At last, the stags have time to chase the roes. And, finally free of worries, the roes allow themselves to be courted; the birth pangs only come two seasons later. Even on the ground, autumn enjoys itself. What we see with our human eyes as a layer of rotting and decay is, in reality, a banquet in which millions of moulds and bacteria are having the time of their life with the now-superfluous leaves and other organs. In autumn, nature harvests itself. After that, it can finally go to sleep – until it has to wake up again the following spring.

Humans aren't harvesters. We're not reapers, we're sowers. We prefer

In autumn, nature is the land of plenty

preparations to results. After six months of hoeing, planting and watering, you see many garden allotments overflowing with abandoned vegetables. If they do get eaten, you see the same mechanism at work again: most of the time is spent on preparations. Whether you harvest your food yourself or buy it at the shops, collecting ingredients, finding recipes, pre-heating the oven, cooking, laying the table and serving takes five times as long as the actual eating. Books about food aren't called food books but cook books. For many people, what was once a daily duty has now become a leisure pastime, if not a culture. People who cook a lot have more prestige than people who eat a lot. But the biggest difference in appreciation of the before and after comes long after the food has been eaten. Defecating has the lowest status. No poems have been written about it; no newspapers have defecating columns next to cookery columns; you defecate on the sly, as if it's illegal. When did you last see someone relieving himself? If you did, ten to one it was a dog, not his owner. But defecating is no less satisfying than eating. Try not doing it. I have friends who've stopped smoking, friends who've stopped drinking and friends who've stopped being friends, but no one in the world can stop defecating. Faeces may be dirty, but defecating is a delight. You see many people leaving the toilet with a vague smile – if not an expression of outright

Defecating is a creative process

pride – on their face. And rightly so, because relieving yourself is a creative process. Something is created – unlike eating, which is only destructive. It's true that it's not a pretty sight; but nor is eating. Just look across the dinner table: all that grinding of teeth, that drowning of food in saliva, those swishing, sloshing sounds, the eager bouncing up and down of the Adam's apple; not even a dinner party is fun after that. When the meal's finished, you feel uncomfortable. Consider, in contrast, how you feel after having visited the toilet. When you defecate, something falls away from you. It's literally a relief. This makes it a pleasure from another era. The need to relieve yourself of part of yourself harks back to the days of blood-letting and enemas. The notion of what constituted well-being was fundamentally different even into the nineteenth century. If you didn't feel well, there was something evil inside you that had to come out. With blood-letting, the excess blood was drawn off to restore the balance with the three other bodily fluids. If blood wasn't the cause, you took something to make you retch or perspire. An enema induced your body to cleanse itself from behind. Today, medicine works in exactly the opposite way. Blood isn't taken out anymore, it's put in. If someone from the seventeenth century were to see a modern blood transfusion, he'd think it was the donor who was being treated. Today, we think something good should be put in rather than something bad taken out: bran, vitamins, plasma, a needle. Out is out, in is in.

Salvation, autumn, faeces: what is actually fulfilment is treated as decay. Waste. Old people are discriminated against. Old age is a reason to sack people, patronize them, steal their handbags or make fun of them. Worst of all is the stereotyping. The elderly are tolerated only if they meet expectations. Being old isn't a state; it's a role. In pubs, old men must have their stock of jokes ready; outside, in their allotments, they're supposed to be leaning over their shovels, with a faraway look in their eyes. Little old ladies are meant not to interfere so much, to think of their incontinence pads on time and to let young people who are in a hurry precede them in the queue. What old people are basically supposed to do is to get out of the way – become invisible, not frighten young people with their existence. Homes for the elderly are built where the ground is cheap, far away from the yuppie apartments in the city centre. Many young people often don't meet an old person for days on end. How are they ever supposed to learn that it was better in the old days?

Old people are increasingly excluded from the fun. Whereas in 1980, half of the men in the Netherlands between the ages of 60 and 65 were still working, today it's less than a third. Not only in the construction industry and on the high seas, but also in government and at universities, there's a decreasing need for older people. You're no longer allowed to work as a professor until you're 70. Your intellectual faculties are supposed to decline dramatically after you turn 65. And discrimination at the workplace isn't the worst of it. You also see fewer and fewer older people at parties, in brothels and near football fields. While a young person can immediately live out every physical desire that wells up in him, if old men or women so much as think about it, they're immediately classified as dirty old folk. Old people live in a world they're no longer part of. "It is the tendency of every society to live and to go on living," wrote Simone de Beauvoir, "it extols the strengths and the fecundity that are so closely linked with use and it dreads the worn-out sterility, the decrepitude of age."

> If old people show the same desires, the same feelings and the same requirements as the young, the world looks upon them with disgust: in them love and jealousy seem revolting or absurd, sexuality repulsive and violence ludicrous. They are required to be a standing example of all the virtues. Above all, they are called on to display serenity; the world asserts that they possess it, and this assertion allows the world to ignore their unhappiness. The purified image of themselves that society offers the aged is that of the white-haired and venerable sage,

rich in experience, planing high above the common state of mankind;
if they vary from this then they fall below it; the counterpart of the first
image is that of the old fool in his dotage, a laughing-stock for his
children. In any case, either by their virtue or by their degradation, they
stand outside humanity.

Simone de Beauvoir wrote this in 1977. But the low status of old age is
much older than that. According to Georges Minois' *History of Old Age*,
"From antiquity to the Renaissance, however societies evolved, they remained
fundamentally based on physical strength and bodily vigour; the conditions
were unfavourable to old age from the start." Only in Roman times, when a
strong State enforced law and order, were the weaker – including the elderly –
protected to some degree from the stronger. Under the ancient Greeks, old
people were worse off than ever. In those days, the perfect body was glorified
as it never has been since – until today. There was no regard at all for wrinkled
faces. According to Aristotle, experience couldn't compensate for decay. In
his eyes, young people had too little experience but old people too much.
Because they'd been wronged too often and made too many mistakes, old
people were insecure and distrustful. Made wise by bad experience, they kept
a tight hand on their purse, and preferred being cowardly to being courageous.
These bad qualities weren't compensated for by the optimism, power and
warmth of youth. If old people loved at all, it wasn't wholeheartedly. They
were better at complaining than laughing, and they were less interested in
the good than the useful. All in all, they lacked faith in the future.

Little has changed. Old people still don't like being old. Life is shrinking.
The nest is emptying, if not the bed, or the head. At the age of 73, the eminent
Dutch writer and Slavic scholar Karel van het Reve complained that he no
longer knew whether he "is or was a writer". His wit was crumbling into
banality. His memory was failing him.

Something occurs to me, a certain passage or a line of argument.
I badly want to write it down. It's a reasoning without any clear gram-
matical form, half intuitive, still half hidden. At the very moment I

feel excited about my new idea and think "I can do something with that", I feel it slip away. It evaporates. That's tragic. Nowadays it happens in the middle of a conversation; just when I'm about to say something very clever about Goethe, the whole name of Goethe escapes me. That's when I think: the next time you want to say anything interesting, please don't open your mouth.

Until his death early in 1999, van het Reve avoided the public eye. He had hoped that when his time came, a suicide pill would be available that could be taken at the moment of one's choice. But, "it mustn't be too expensive, because I'm too stingy for that. I don't intend to pay more than 20 or 30 guilders for it."

Outsiders and initiates thus agree on what old people look like and what they are: excluded, a burden on society, not worthy of death and long-time stinges. Stereotypes underlie all forms of discrimination. But how can you ever get rid of them if the discriminated themselves agree so readily with the stereotypes? If someone considers himself an old fogey, how can he expect the outside world to think otherwise? Fortunately for commerce, old people have poor memories. If you don't remind them they're old, they forget it. Advertising agencies learned this lesson the hard way. You may sell a lot to young people by putting "YOUNG!" on the label, but you won't sell a thing to old people by putting "OLD!" on it. That's too bad, because old people have lots of money and oceans of time to spend it in. And there are more of them every day, getting older by the minute. But they don't want to read a newspaper for the elderly, any more than an advertiser wants to take the risk of his product being seen as something for old people. The old do like magazines about nature, though, and radios with as few knobs as possible and packaging that's easy to open. So, advertise with promises of "nature", "convenience" and – especially – "health", because older people will respond to that. And as it turns out, a lot of young people will end up responding to it too. You're never too young for nature, convenience and health. This is how you break open young markets with old people.

What else are old people good for? To find out, you have to look at how old people differ from young people. Take the Holy Trinity. The hero of the three is Jesus Christ, into whom the Holy Ghost breathed life. As becomes a hero, he didn't grow old: he died at the age of 33, long before decline could set in. But, unlike the Roman and Berlin Empires, the Jesus Empire continues to exist. Sure enough, the Young Hero was helped by His Elderly Father. Just

Charles Darwin at the age of 81

as Jesus Christ is invariably depicted as a Hollywood star with a short beard – it never takes long to find a leading man for Bible films – God the Father is always an old man with a long beard, the quintessential Santa Claus. Long beards are the symbol of both old age and wisdom. Patriarchs, prophets and missionaries all have them. Beards cover a large part of the face, giving it a mysterious air. More importantly, they obscure the lower part of the face. Just as the lower half of the body stands for man's animal side, the lower half of the face – the lips, the jaws, the teeth, the mucus – represents his lower instincts. On the other hand, beards accentuate the eyes and the forehead, as the spiritual, loftier parts. Wisdom, like long beards, comes with age, you can see that straight away. While the body declines, the spirit grows.

In the past, in the days before the Internet, British Telecom and computers, old people formed a reservoir of wisdom from which the entire grateful community could benefit and of which it gladly made use. In *Onder professoren* (Among Professors), Dutch novelist W. F. Hermans has one of his characters summarize this notion:

A hundred years ago it was a great privilege to be a healthy old man. Old people were revered because they were scarce. They were listened to because they had experience. And it was worth listening to them, because their experience was, indeed, valuable on occasion. You were happy if you were old. Becoming old was a reward for a life full of misery. But today? The only things considered important today are those things reserved for young people: revolutions, cycling, pop music, boxing, flying to the moon in a rocket, sex. And everything changes so quickly! There you are with your obsolete experience! Oh God . . . slaved away for years to learn something and then it's not needed for anything anymore. Old people – they're the Jews and gypsies of the future.

It was better in the past. Today, the past is better only in those countries where it's still the past. Indian chiefs, tribal elders, powerful sheikhs, prelates: the older they were the better. That's what Rudyard Kipling, Tintin and Hollywood have taught us. But wise is something else – what counts here isn't their age, but our sense of romance.

A sense of romance means believing that somewhere else in the past it was better – or that somewhere else in the future it will be better. Indonesians who have lived since the colonial days in The Hague, where old folk are abandoned in homes for the elderly, nostalgically tell their grandchildren how well old people are listened to back home. They're deeply disappointed when they return there for a visit. It's true, old people do get treated with respect, but they're not so old after all. The minds of the elderly are listened to only as long as their bodies are in order. When the old men become infirm, they're dropped from the village's ruling council in just as subtle or brutal a way as the president of the stamp collectors' society who can no longer read his own signature. This certainly doesn't apply only to Indonesians. Studies have shown that more than half of the non-Western societies treat their old folk badly. They're underfed and poorly cared for when they're ill. But then, aren't they supposed to be nearer the world of the spirits, and to put in a good word for us there? Perhaps. Connections with the world of the spirits also have their drawbacks though. In many cultures, old women and men are made out to be witches. How else could they have grown to be so old?

Wisdom responds to the laws of supply and demand. Old people were

In many cultures, old women are seen as witches

safe in societies whose survival was based on oral traditions. They were essential as a channel through which to pass things down from one generation to the next. Without this collective memory, one wasn't allowed to administer justice in ancient Greece or medieval Europe. Old people's worst enemies are books. Experiences can be registered in books as well as in old folk, and libraries are cheaper than homes for the elderly. In our technological age, the knowledge stored in old people pales in the face of all those computer files. Moreover, our whole society revolves around change. In this context experience can be a burden. To learn something new, one first has to unlearn something old. So learning is twice as much work for an old person as it is for a person with no experience. In a study conducted in Berlin, a group of old people had to learn a new skill, for which they first had to acquire new knowledge. It wasn't easy. No wonder old people have problems with new equipment – all those switches and fancy gadgets. Our society moves too quickly for old people. But maybe that's handy because then something can be done about it sooner. Just wait until the baby-boomers become pensioners. With that same flair that brought "young" to power during their youth, they'll soon make tomorrow's young people understand in no uncertain terms that, unfortunately for them, "old" is now in fashion.

Old people probably had the highest status when they were scarcest. It was long before the invention of writing that they were needed most and died fastest. In those days, human lives were as short as those of the other apes. The survival of apes depends as much on knowledge as ours does; more knowledge than can be encoded in their genes at birth. They have to learn a lot from the older members of the group. Once mature, they allow themselves to be led by an experienced male; usually he's one of the oldest members. What he lacks in power is compensated for in wisdom. Even if he's too old actually to lead, his advice is still followed. In a chimpanzee colony at the Arnhem Zoo in the Netherlands, there was even an old female – Mama – who, with her great wisdom, had considerable influence on the well-being of the troupe. Precisely because she had no official status, which could have been contested, she continually brought peace and stability to the group. According to the well-known Dutch researcher Frans de Waal, her central position "was comparable to that of a grandmother in a Spanish or Chinese family". When the tensions in the group reached unbearable levels, the opposing parties, including adult males, always came to her. Fights between two males repeatedly ended in her embrace. However, the chimpanzee world isn't idyllic either. When chimpanzees become sickly and really start to fail, the others literally leave them to drop dead.

Mama, the wise old chimpanzee

Elephants are an exception. It turns out they have respect not only for death but also for age. The oldest female member knows better than any other member of the herd where water can be found in the dry season and where the best escape routes are. The whole herd follows such a matriarch. She doesn't get dumped when she starts to decline. Quite the opposite. A herd in Africa is known to have been led by a blind sexagenarian. Assisted by helpers, she showed them the migratory paths using her sense of smell. When she finally died, the herd went to pieces for weeks.

The only chance an animal has of growing old is within a group. The more social the species, the greater the chance. Canines care as well for the older generation as they do for the younger. The stay-at-homes wait until the hunting party returns and regurgitates the food – first for the young, then for the old and finally for the pup minders. Tigers needn't count on members of their species. Their character is too much that of the loner. The best a tiger can do when it becomes decrepit is to concentrate on the easiest prey for a while and become a man-eater. In the past, in India the life of one old, female tiger was prolonged at the expense of dozens of villagers. In this age of modern weapons, you don't even grow old as a man-eater anymore. To find old tigers you have to go to the zoo. Now that breeding in captivity is becoming increasingly successful and fewer and fewer animals are being removed from the wild, many zoos are gradually becoming more and more like homes for the elderly. That many homes for the elderly resemble zoos has another explanation.

Old apes, monkeys and elephants are tolerated because they have knowledge the group needs. When antelopes are old, the only thing many of them can contribute to the group is their body: the herd drives them to the edges, where they serve as decoys for predators, who are out to get the weakest prey anyway. As far as the herd is concerned, they can have the old ones, but the vulnerable young ones are strictly off limits. While the predator absconds with the old peace offering, the young are guarded like valuable treasures in the bosom of the group. Honey-bees don't need a helping hand; they sacrifice themselves. During their first weeks, the worker bees do the housekeeping in the hive. They feed the larvae and keep everything shipshape. Then, after about three weeks, they're allowed to go shopping. As the weeks pass, they quickly wear out from all the work. The more conscientious the worker bees are, the weaker they become and the sooner they fall prey to spiders or other bee-eaters. Danger lurks everywhere outside, but there's not a fellow bee that minds: there are bees aplenty. As long as the young worker bees nurture the larvae into new, young worker bees, the old worker bees can literally work themselves to death. This takes a little getting used to for us humans, who send precisely our young men out into battle. But if you go and sit on an anthill, it's the oldest ants that raise their abdomens to inject the poison into your backside. You risk your life if there's not much left of it to live. Incidentally, the biggest difference between their six-legged soldiers and our two-legged ones isn't even that it is the old bees and ants doing the dirty work, but that it's the females. Among ants, the females are the most ferocious fighters. And the older the female, the more aggressive she is. But that sometimes seems to be the case among humans too.

Biologically, an old person is poorly off. No one wants his body anymore, and his knowledge is only good for the museum. He's at the mercy of his fellow man. It's that simple. Who, in our society, still takes their old mother into their home? Today, to be allowed to spend your last days in our homes, you have to be a cat or a dog. A survey in the Netherlands revealed that 29 per cent of Dutch dogs and 16 per cent of Dutch cats are more than ten years old. Old pets aren't disposed of like old mothers. People stay faithful to their animals till death do them part. Homes for elderly cats and dogs don't even exist. There are shelters for stray and asylum-seeking pets, but old cats and dogs don't need to be taken in. While old people, driven out of hearth and home, sit silently in the social room until death doth follow, allowing their backsides to be washed by impudent young brats with rings in their ears whom they've never seen before, venerable old cats purr away on their own cushions next to the family hearth that's been stoked a bit warmer just for them. There are also rest homes for old horses and cows, but they've already lived in communal quarters all their lives. As noble beasts, horses were the

A Dutch farmer and his wife say farewell to
their beloved horse Corrie, 1962

first animals in the Netherlands to be given rest homes. Retirement homes for cows are found mostly in India, where it's a sacred duty not to let them suffer, but recently, a refuge for cows was set up in the Netherlands. A normal dairy cow there goes to the slaughterhouse after about five years, but the select few, like Claartje, are allowed to live out what's left of their 20 or 30 years in this refuge. Next to Claartje stands a cow who's taking it easy after producing 120,000 litres of milk during her long lifetime. Now, she can relax and enjoy herself. She doesn't have to do anything, not even make milk. It's the same story in stables, with that one horse that's always allowed to linger on: a record of all the carts he pulled, the medals he won, the little girls he gave rides to. This is anthropomorphism in the extreme. Surely old people, too, have more than earned a peaceful dotage. To be allowed to do nothing, one first has to have worked hard. First bent from working; then bent from rheumatism. "Enjoy it," says the nurse.

Most old people don't enjoy enjoying. They prefer complaining. After all those years, they're experts at it. When they've finally finished their complaining, old women do what women often do when they've finished complaining: put on comfortable shoes so that they can go into town to buy uncomfortable shoes, shoes that don't make them look so old. Unfortunately, while they're in town they seldom look around them: with a bit of luck, the city centre is full of old buildings, some of them even older than they are. That in itself is a comfort. Why is all that old stuff still there? There might have been a new housing development instead, or a marble-tiled shopping plaza, a shady park, nightclubs. Yet people prefer to see their old pubs, their old rectories, their old gatehouses. They're proud of them. Proud of piles of stones. And why? Because, like sponges, they've absorbed history. People need history, even if only to be able to distance themselves from the present. One should – to echo Ivan Illich – be able to see the present as the future of the past. An old city offers the perfect opportunity. Its houses have survived wars, watched children grow up, taken people into hiding, witnessed parties, stored tea crates. Luckily, one can still see this. Old houses radiate something that old people today rarely exude: venerability.

The nice thing about venerability is that it increases with age. It needs time. Venerability is what a city hall has but a city office doesn't; it's the difference between Mother Goose and Donald Duck, Queen Victoria and the Prince of Wales. He who tries to incorporate it into his buildings from the very beginning gets stuck in the megalomaniac constructions of a Mussolini or communist architecture. Patience is required. A few old columns on a Greek hill are more impressive than all of the spanking new Docklands. The tourists are attracted by it, even though they don't know why. Venerability is victory over time. It's the opposite of decay; it's fulfilment.

My grandfather was good at it. While some old people look young, he was young and looked old – and it suited him. Impeccably dressed, a summer hat in summer, a winter hat in winter, complete with gaiters and – to the great delight of his grandchild – always a cigar cutter on his watch chain. A miniature guillotine that sparked my imagination from the first time I saw it, the cutter was often used, because a real grandfather smokes cigars. A grandfather who doesn't doesn't know what being a grandfather is all about. You're either 70 or you aren't. Each age has its role: babies are supposed to fill nappies, adolescents to balance their suicidal tendencies with indecision, young men to go through hell and high water for ideals they'll later come to regret . . . And old people are supposed to be old. It's childish to want to be a different age than you are. Children want to be grown up later. Be glad if it happens. If someone who is 70 constantly tries to look 50, how can someone who is 30 muster respect for someone who's 70?

You'll never become venerable that way. It has to do with self-respect.

People who want to stay young don't prolong their lives; they shorten them. They miss the second half of life's stairway. Because the panorama was so beautiful on the way up, they don't dare look down during the descent. It's like not daring to return to the theatre after the intermission for fear the play will end. But that's precisely the point of plays, films and novels: they end. The more beautiful the story, the more you wish it would never end, and, at the same time, the more you can't wait to get to the end. How can something interest you if you're not curious about how it ends? What could be more exciting than to know how your life will turn out? As a boy, I always wanted to know what it would be like to be a girl. Not out of dissatisfaction with myself, but because it's fun to immerse yourself in someone else. But it'll never go beyond an idea; I will never personally experience what it's like to be a woman. I can, though, still become someone else. I can become an old man. I look forward to it, because I've never been an old man before. Only then will I really become myself. As one person put it, "the older you become, the more you come to resemble yourself." Everything you experience in life leaves its mark on you – different marks on different people to different degrees. With every step you take in the direction of the grave, you look less like another and more like yourself – until, in the end, you become one with yourself. Then, there's nothing left to be done. It is finished.

BIBLIOGRAPHY

Aafjes, Bertus, *De karavaan*. Amsterdam: De
Ceder, 1953.

Allen, Judy and Jeanne Griffiths. *The Book of the
Dragon*. London: Orbis, 1979.

Abeles, Ronald P., Helen C. Gift and Marcia G.
Ory, eds. *Aging and Quality of Life*. New York:
Springer, 1994.

Alberti, Leon Battista. *On the Art of Building in Ten
Books*. Translated by Joseph Rykwert et al.
Cambridge (Mass.): Harvard University
Press, 1988.

Andrews, Carol. *Egyptian Mummies*. London:
British Museum Publications, 1984.

Andrew, Warren. *The Anatomy of Aging in
Man and Animals*. New York: Grune
& Stratton, 1971.

Andrews, Michael. *The Life that Lives on Man*.
London: Faber & Faber, 1976.

Arber, Sara and Jay Ginn. *Gender and Later Life*.
London: Sage, 1991.

Ariès, Philippe. *Studien zur Geschichte des Todes im
Abendland*. Munich: Hanser, 1976.

Ariès, Philippe. *Western Attitudes to Death from the
Middle Ages to the Present*. Translated from the
French by Patricia M. Ranum. London:
Marion Boyars, 1976.

Ariès, Philippe. *The Hour of Our Death*. Translated
from the French by Helen Weaver. London:
Allen Lane, 1981.

Ariès, Philippe. *Het uur van onze dood: duizend jaar
sterven, begraven, rouwen en gedenken*.
Amsterdam and Brussels: Elsevier, 1987.

Aristotle. "*De longitudine et brevitae vitae*",
translated by G. R. T. Ross. In *The Works of
Aristotle*, translated into English under the
editorship of J. A. Smith and W. D. Ross.
Oxford: Clarendon Press, 1908–1952.

Arking, Robert. *Biology of Aging: Observations
and Principles*. Englewood Cliffs (N.J.):
Prentice-Hall, 1991.

Barbier, Patrick. *Histoire de castrats*. Paris:
Grasset, 1989.

Bartels, D. *Ambon is op Schiphol: socialisatie,*

*identiteitsontwikkeling en emancipatie bij
Molukkers in Nederland*. Leiden and Utrecht:
Centrum voor Onderzoek van
Maatschappelijke Tegenstellingen en
Inspraakorgaan Welzijn Molukkers, 1990.

Barthélemy, Guy. *Les Jardiniers du Roy: une petite
histoire du Jardin des Plantes de Paris*. Paris:
Le Pélican, 1979.

Beaufort, C. C. Th. de. *Het behouden van bouww-
erken uit de oudheid*. Haarlem: De erven F.
Bohn, 1969.

Beauvoir, Simone de. *The Coming of Age*. New
York: G. P. Putnam & Sons, 1972.

Beauvior, Simone de. *Old Age*. Translated by
Patrick O'Brian. Harmondsworth:
Penguin, 1977.

Beek, James and Michael Daley. *Art Restoration:
The Culture, the Business and the Scandal*.
London: John Murray, 1993.

Benjaminse, M. E. and I. C. Laboyrie. *De
wasmodellen van Petrus Koning*. Utrecht:
Universiteitsmuseum, 1985.

Berenbaum, May R. *Bugs in the System: Insects and
their Impact on Human Affairs*. Reading:
Addison-Wesley, 1995.

Berends, Rob. *Monumentenwijzer*. The Hague:
Sdu, 1995.

Berg, Arie van den. *Van binnen moet je wezen*.
Amsterdam: De Arbeiderspers, 1989.

Bergeon, S. "*Science et patience*" *ou la restauration
des peintures*. Paris: Réunion des Musées
Nationaux, 1990.

Bergvelt, Ellinoor and Renée Kistemaker, eds.
*De wereld binnen handbereik: Nederlandse kunst –
en rariteitenverzamelingen, 1585–1735*. Zwolle
and Amsterdam: Waanders and Amsterdams
Historisch Museum, 1992.

Berk, Marjan. *Koken met kraaiepoten*. Wormer:
Inmerc, 1993.

Bertin, Léon et al. *Buffon*. Paris: Le Muséum
National d'Histoire Naturelle, 1952.

Blunt, Wilfrid. *The Compleat Naturalist: A Life of
Linnaeus*. London: Collins, 1971.

Boeke, J. *Problemen der onsterfelijkheid: leven, dood, levensduur en onsterfelijkheid, biologisch beschouwd.* Amsterdam: J. M. Meulenhoff, 1941.

Boheemen, Petra van and Paul Dirkse. *Duivels en demonen: de duivel in de Nederlandse beeldcultuur.* Utrecht: Museum Het Catharijneconvent, 1994.

Bonner, John Tyler. *Life Cycles: Reflections of an Evolutionary Biologist.* Princeton (N.J.): Princeton University Press, 1993.

Boonen, Christianne. *Een steenhard bestaan: zeldzame muurbegroeiing langs Utrechts grachten.* Utrecht: Westers, 1982.

Bowden, Douglas M., ed. *Aging in Nonhuman Primates.* New York: Van Nostrand Reinhold, 1979.

Bradbeer, J. W. *Seed Dormancy and Germination.* Glasgow: Chapman & Hall, 1988.

Brand, Stewart. *How Buildings Learn: What Happens After They Are Built.* Harmondsworth: Penguin, 1994.

Brink, R. H. S. van den. *Attitude and Illness Behaviour in Hearing-Impaired Elderly.* Proefschrift Groningen, 1995.

Brodie, Harold J. *Fungi: Delight of Curiosity.* Toronto, Buffalo and London: University of Toronto Press, 1978.

Bronswijk, Johanna E. M. H. van. *House Dust Biology for Allergists, Acarologists and Mycologists.* Zeist: NIB, 1981.

Brookbank, John W. *The Biology of Aging.* New York: Harper & Row, 1990.

Brophy, John. *The Human Face Reconsidered.* London: George G. Harrup & Co, 1962.

Brothwell, Don. *The Bog Man and the Archaeology of People.* Cambridge (Mass.): Harvard University Press, 1987.

Burnet, F. M. *Intrinsic Mutagenesis: A Genetic Approach to Aging.* New York: J. Wiley & Sons, 1974.

Burton, Maurice. *Living Fossils.* London and New York: Thames and Hudson, 1956.

Buss, Leo W. *The Evolution of Individuality.* Princeton (N.J.): Princeton University Press, 1987.

Bustad, Leo K. *Animals, Aging and the Aged.* Minneapolis: University of Minnesota Press, 1980.

Calder, W.A. *Size, Function and Life History.* Cambridge (Mass.): Harvard University Press, 1984.

Camporesi, Piero. *The Incorruptible Flesh: Bodily Mutilation and Mortification in Religion and Folklore.* Translated from the Italian by Tania Croft-Murray. Cambridge: Cambridge University Press, 1988.

Cantor, Norman. *Medieval History: The Life and Death of a Civilisation.* New York: MacMillan, 1963.

Chapman, Philip. *Caves and Cave Life.* London: HarperCollins, 1993.

Chargesheimer. *Schöne Ruinen.* Cologne: Wienand, 1994.

Charlesworth, B. *Evolution in Age: Structured Populations.* Cambridge: Cambridge University Press, 1980.

Child, C. M. *Senescence and Rejuvenenscence.* Chicago: University of Chicago Press, 1915.

Christensen, Clyde M. *The Molds and Man.* 1951. Minneapolis: University of Minnesota Press, 1972.

Cobb, Gerald. *English Cathedrals: The Forgotten Centuries.* London: Thames & Hudson, 1980.

Cole, Thomas R. and Mary G. Winkler, eds. *The Oxford Book of Ageing.* Oxford and New York: Oxford University Press, 1994.

Comfort, Alex. *The Biology of Senescence.* Edinburgh: Churchill Livingstone, 1979.

Cool, Catherina and van der Lek. *Paddenstoelenboek.* 1913. Amsterdam, Batavia and Paramaribo: W. Versluys, 1936.

Cornaro, Luigi. *The Art of Living Long.* Milwaukee: William F. Butler, 1918.

Cottrell, Leonard. *Lost Cities.* London: Robert Hale, 1957.

Curtis, Howard. J. *Biological Mechanisms of Aging.* Springfield (III.): Charles C. Thomas, 1966.

Dante Alighieri. *De goddelijke komedie (The Divine Comedy).* Dutch translation from the Italian by Frans van Dooren. Baarn and Amsterdam: Ambo and Athenaeum-Polak & Van Gennep, 1987.

Darlington, A. *Ecology of Walls*. London: Heinemann Educational Books, 1981.

David, Rosalie and Eddie Tapp. *Evidence Embalmed: Modern Medicine and the Mummies of Ancient Egypt*. Manchester: University Press, 1984.

Davidson, Gustav: *A Dictionary of Angels: Including the Fallen Angels*. New York: The Free Press, London: Collier-MacMillan, 1967.

Davis, Kenneth S. and John A. Day. *Water: The Mirror of Science*. London: Heinemann, 1964.

Davis, Kathy. *Reshaping the Female Body: The Dilemma of Cosmetic Surgery*. London: Routledge, 1995.

Dawkins, Richard. *The Blind Watchmaker*. London and New York: W.W. Norton & Company, 1986.

Dekkers, Midas. *Bestiarium*. Amsterdam: Bert Bakker, 1977.

Dekkers, Midas and Han van Hagen. *Gekorven diertjes*. Heusden: Aldus, 1983.

Dekkers, Midas and Han van Hagen. *Mummies*. Amsterdam: Contact, 1986.

Dekkers, Midas. *Cahier ECOS: Mens, energie & milieu*. The Hague: Museon/Ministerie VROM, 1987.

Dekkers, Midas. *Dearest Pet: On Bestiality*. Translated from the Dutch by Paul Vincent. London: Verso, 1994.

Denslagen, W.F. *Omstreden herstel: kritiek op het restaureren van monumenten*. The Hague: Staatsuitgeverij, 1987.

Deth, Louise van and Johanne Radersma. *Dwars door de overgang*. Amsterdam: Meulenhoff, 1995.

DiGiovanna, Augustine Gaspar. *Human Aging: Biological Perspectives*. London and New York: McGraw-Hill, 1994.

Douglas, Mary. *Purity and Danger: An Analysis of Concepts of Pollution and Taboo*. 1st ed. 1966. London: Routledge, Kegan Paul, 1978.

Draaisma, Douwe. *Het verborgen raderwerk: Over tijd, machines en bewustzijn*. Baarn: Ambo, 1990.

Drewermann, Eugen. *Over de onsterfelijkheid van de dieren: Hoop voor het lijdende schepsel*. Amsterdam: De Driehoek, 1993.

Drimmer, Frederick. *Very Special People: The Struggles, Loves and Triumphs of Human Oddities*. New York: Amjon, 1973.

Dunning, A.J. *Uitersten: beschouwingen over menselijk gedrag*. Amsterdam and Utrecht: Meulenhoff and Wetenschappelijke Uitgeverij, 1990. Published in English as *Extremes: Reflections on Human Behaviour*. Translated by Johan Theron. London: Secker and Warburg, 1993.

Evans, Howard Ensign. *Life on a Little-Known Planet*. Chicago and London: University of Chicago Press, 1968.

Everdingen, J. J. E. van and N.S. Klazinga. *Maat en getal in taal en teken: geneeskundig gestoei met cijfers en letters*. Utrecht and Antwerp: Bohn, Scheltema & Holkema, 1988.

Everdingen, J. J. E. van, ed. *Beesten van mensen: microben en macroben als intieme vijanden*. Overveen: Belvedere, 1992.

Ex, Nicole. *Zo goed als oud: de achterkant van het restaureren*. Amsterdam: Amber, 1993.

Faas, P. C. P. *Rond de tafel der Romeinen, met meer dan 150 originele recepten*. Diemen: DOMUS, 1994.

Farson, Daniel and Angus Hall. *Mysterious Monsters*. London: Aldus, 1978.

Feltkamp, C. *De begrafenismoeilijkheden in 1945 te Amsterdam*. Amsterdam: Bureau voor Pers, Propaganda en Vreemdelingenverkeer, 1978

Fenema, Joyce van. *Wat dacht je van een nieuw lijf? – alles over esthetische plastische chirurgie*. The Hague: BZZTôH 1995.

Ferraro, K. F., ed. *Gerontology: Perspectives and Issues*. New York: Springer, 1990.

Fiedler, Leslie. *Freaks: Myths and Images of the Secret Self*. New York: Simon and Schuster, 1978.

Finch, Caleb E. *Longevity, Senescence and the Genome*. Chicago and London: University of Chicago Press, 1990.

Franssen, Maarten. *Archimedes in bad: mythen en sagen uit de geschiedenis van de wetenschap*. Amsterdam: Prometheus, 1990.

Frisch, Karl von. *Zehn kleine Hausgenossen*. Munich: Ernst Heimeran, 1940. Published in English as *Ten Little Housemates*. Translated by

Margarte D. Senft. Oxford: Pergamon Press, 1960.

Garfield, Sydney. *Teeth Teeth Teeth: A Treatise on Teeth and Related Parts of Man, Land and Water Animals.* New York: Simon and Schuster, 1971.

Godwin, Macolm. *Angels.* London: Boxtree, 1993.

Gordon, David George. *The Compleat Cockroach.* Berkeley (Cal.): Ten Speed Press, 1996.

Goudsmit, Samuel A. and Robert Claiborne. *Time.* New York: Time Books, 1966.

Gould, Stephen Jay. *The Panda's Thumb: More Reflections in Natural History.* London and New York: W. W. Norton & Company, 1980.

Gould, Stephen Jay. *The Mismeasure of Man.* London and New York: W. W. Norton & Company, 1981.

Gould, Stephen Jay. *Hen's Teeth and Horse's Toes.* London and New York: W. W. Norton & Company, 1983.

Gould, Stephen Jay. *Time's Arrow, Time's Cycle: Myth and Metaphor in the Discovery of Geological Time.* Cambridge (Mass.) and London: Harvard University Press, 1987.

Gould, Stephen Jay. *Life's Grandeur: The Spread of Excellence from Plato to Darwin.* London: Jonathan Cape, 1996. Published in the U.S.A. as *Full House.* New York: Harmony Books, 1996.

Green, Patricia Dale. *Cult of the Cat.* New York: Weathervane, 1963.

Grijp, Louis Peter, Everdien Hock and Annemies Tamboer, eds. *De dodendans in de kunsten.* Utrecht: HES, 1989.

Haan, Hilde de and Ids Haagsma, *Architects in Competition: International Architectural Competitions of the Last 200 Years.* London: Thames and Hudson, 1988.

Habermehl, Karl-Heinz. *Die Altersbestimmung bei Haus und Labortieren.* Berlin and Hamburg: Paul Parey, 1975.

Haire, Norman. *Rejuvenation: The Work of Steinach, Voronoff and Others.* London: George Allen and Unwin, 1924.

Haldane, J. B. S. *On Being the Right Size and Other Essays.* Oxford and New York: Oxford University Press, 1985.

Hamilton, David. *The Monkey Gland Affair.* London: Chatto & Windus, 1986.

Haneveld, G. T. *Het mirakel van het hart.* Baarn: Ambo, 1991.

Hapgood, Fred. *Why Males Exist: An Enquiry into the Evolution of Sex.* New York: William Morrow & Co., 1979.

Hart, Maarten 't. *Ratten: over het gedrag, de leefwijze en het levermogen van de rat, over de rattenbestrijding en de rattenkoning. Met enkele aanwijzingen voor het houden van de rat als huisdier.* Amsterdam: Wetenschappelijke Uitgeverij, 1973.

Hartnack, Hugo. *Unbidden House Guests.* Tacoma (Wash.): Hartnack, 1943.

Hawking, Stephen. *A Brief History of Time.* London, Bantam, 1998.

Hayflick, Leonard. *How and Why We Age.* New York: Ballantine Books, 1994.

Hazelzet, Korine. *Heethoofden, misbaksels en halve garen: de bakker van Eeklo en de burgermoraal.* Zwolle: Catena, 1988.

Hazelzet, Korine. *De levenstrap.* Zwolle: Catena, 1988.

Hellema, Henk. *Geur en gedrag.* Amsterdam: De Brink, 1994.

Henderson, Michael. *The BMA Guide to Living with Risk.* New York: John Wiley & Sons, 1987.

Hendrickson, Robert. *More Cunning than Man: A Social History of Rats and Men.* New York: Dorset Press, 1983.

Hepkema, Jacob. *Wieuwerd en zijn historie.* 1st ed. 1896. Oosterend: Van der Eeems, 1977.

Hermans, W. F. *Onder Professoren.* Amsterdam: De Bezige Bij, 1975.

Herodotos. *Historiën.* Dutch translation from the Greek by Onno Damste. Bussum: De Haan, 1968.

Hickin, N. E. *Termites: A World Problem.* London: Hutchinson, 1971.

Hickin, Norman. *Bookworms: The Insect Pests of Books.* London: Sheppard Press, 1985.

Hillenius, D. *Het romantisch mechaniek.* Amsterdam: De Arbeiderspers, 1969.

Hillenius, D. *De hand van de slordige tuinman.* Amsterdam: Oorschot, 1996.

Howell, Michael and Peter Ford. *The True History of the Elephant Man.* London: Allison & Busby, 1980.

Jackson, John Brinckerhoff. *The Necessity for Ruins and Other Topics*. Amhurst: The University of Massachusetts Press, 1980.

Jacobs, Jane. *The Death and Life of Great American Cities*. 1st ed. 1961. Harmondsworth: Penguin, 1994.

James, J. *Celveroudering en celdood*. Amsterdam: Amsterdam University Press, 1994.

Johnson, Julia and Robert Salter, eds. *Ageing and Later Life*. London, Thousand Oaks and New Delhi: Sage Publications, 1993.

Jones, Steve. *The Language of the Genes: Biology, History and the Evolutionary Future*. London: HarperCollins, 1993.

Jones, Steve: *In the Blood: God, Genes and Destiny*. London: HarperCollins, 1996.

Joost, Th. Van and L. Reijnders. *Milieu en huid: de huid als spiegel van het milieu*. Meppel and Amsterdam: Boom, 1992.

Joost, Th. van and J. J. E. van Everdingen. *Meer dan huid alleen: cultuurhistorische verkenningen*. Amsterdam and Overveen: Boom en Belvédère, 1996.

Knipping, John B. *Pieter Bruegel de Oude: de val der opstandige engelen*. Leiden: L. Stafleu, 1949.

Knutson, Roger, M. *Furtive Fauna: A Field Guide to the Creatures who Live on You*. New York: Penguin Books, 1992.

Koch, Tankred. *Levend begraven*. Baarn: Bigot & Van Rossum, 1992.

Köhler, Wim. *Lang en gelukkig?: levensverwachting en doodsoorzaken van Nederlanders*. Utrecht and Antwerp: Kosmos, 1992.

Kohn, Robert R. *Principles of Mammalian Aging*. Englewood Cliffs (N.J.): Prentice Hall, 1978.

Köster-Lösche, Kari. *Die sieben Todesseuchen, von Pest bis AIDS von Altertum bis heute*. Husum and Nordsee: Cobra, 1989.

Kousbroek, Rudy. *De onmogelijke liefde*. Amsterdam: Meulenhoff, 1988.

Kreienbühl, Jürg. *Le monde merveilleux de la Galerie de Zoologie*. Basle: Galerie Specht, 1988.

Kruit, Wilfred and Govert Schilling. *Dimensies in de natuur*. Amsterdam: Aramith, 1987.

Kruit, Wilfred. *Leeftijd: een speurtocht naar de biologische oorzaken van veroudering*. Amsterdam and Antwerp: Contact, 1996.

Krutch, Joseph Wood. *The Great Chain of Life*. Boston: Houghton Mifflin, 1956.

Kruyt, W. *Wat groeit en bloeit op oude muren*. Zutphen: Thieme, 1987.

Kurtén, Björn: *How to Deep-Freeze a Mammoth*. New York: Columbia University Press, 1986.

Laermans, Rudi. *Individueel vlees: over lichaamsbeelden*. Amsterdam: De Balie, 1986.

Lam, Ineke. *De mythe van het "lege nest": over echtpaarrelaties als de kinderen het huis uit zijn*. Utrecht: Dissertation, 1994.

Laudau, Terry. *About Faces*. New York : Doubleday, 1989.

Laslett, P. *A Fresh Map of Life: The Emergence of the Third Age*. Cambridge, (Mass.): Harvard University Press, 1991.

Leche, Wilhelm. *Der Mensch: sein Ursprung und seine Entwicklung*. Jena: Gustave Fischer, 1911.

Levy, Mattthys and Mario Salvadori. *Why Buildings Fall Down: How Structures Fail*. London and New York: W. W. Norton & Company, 1994.

Liedekerke, Anne-Claire de and Hans Devisscher, eds. *Fiamminghi a Roma 1508/1608: kunstenaars uit de Nederlanden en het prinsbisdom Luik te Rome tijdens de Renaissance*. Brussels and Ghent: Vereniging voor tentoonstellingen van het Paleis voor Schone Kunsten and Snoeck-Ducaju & Zoon, 1995.

Lodrick, Deryck O. *Sacred Cows, Sacred Places: Origins and Survivals of Animal Homes in India*. Berkeley, Los Angeles and London: University of California Press, 1981.

Lowenthal, D., ed. *Our Past Before Us: Why Do We Save It?* London: Temple Smith, 1981.

Macaulay, Rose. *The Pleasure of Ruins*. Photographs by Roloff Beny. London: Thames and Hudson, 1964.

Mak, Geert. *De Engel van Amsterdam*. Amsterdam: Atlas, 1992.

McDannell, Colleen and Bernhard Lang. *Heaven: A History*. New Haven and London: Yale University Press, 1988.

McGee, Harold. *On Food and Cooking: The Science and Lore of the Kitchen*. 1984. London and Sydney: Unwin Hyman, 1987.

McGrady, Patrick M. *The Youth Doctors*. London:

Arthur Barker, 1969.

McHargue, Georgess. *Mummies: stille getuigen uit het verleden.* Delft: Elmar, 1969.

McMahon, Thomas A. and John Tyler Bonner. *On Size and Life.* New York: Scientific American Library, 1983.

Medawar, Peter B. *Ageing: An Unsolved Problem of Biology.* London: H. K. Lewis & Co., 1952.

Medawar, P. B. *The Uniqueness of the Individual.* London: Methuen & Co., 1957.

Meeuse, Bastiaan and Sean Morris. *The Sex Life of Flowers.* London: Faber, 1984.

Middelkoop, Norbert. *De anatomische les van Dr Deijman.* Amsterdam: Amsterdams Historisch Museum, 1994.

Minios, Georges. *History of Old Age: From Antiquity to the Renaissance.* Cambridge: Polity Press, 1989.

Moore-Ede, M. C., F. M. Sulzmann and Ch. A. Fuller. *The Clocks that Time Us.* Cambridge, (Mass.) and London: Harvard University Press, 1982.

Morgan, Elaine. *The Scars of Evolution.* London: Penguin Books, 1991.

Morowitz, Harold J. *Entropy and The Magic Flute.* New York and Oxford: Oxford University Press, 1993.

Mostafavi, Mohsen and David Leatherbarrow. *On Weathering: The Life of Buildings in Time.* Cambridge (Mass.) and London: The MIT Press, 1993.

Moudon, Anne Vernez. *Built for Change.* Cambridge (Mass.): The MIT Press, 1986.

Nater, J. P. *De dood is in de pot, man gods!: ziekte en genezing in de bijbel.* Rotterdam: Erasmus Publishing, 1994.

Nesse, Randolph M. and George C. Williams. *Why We Get Sick: The New Science of Darwinian Medicine.* New York: Time Books, 1994.

Niesthoven, J. C. *Informatie in woord en beeld over schadelijke en lastige dieren in en om het huis.* Amsterdam: Moussault's Uitgeverij, 1970.

Niesthoven, J. C. *Enge beestjes in huis: over ratten en muizen, faramieren, spinnen . . .* Amsterdam and Antwerp: Kosmos, 1979.

Nooden, L. D. and A. C. Leopold. *Senescence and Aging in Plants.* New York: Academic Press, 1988.

Nijhof, Peter, et al. *Langs pakhuizen, fabrieken en watertorens: industrial-archeologische routes in Nederland en België.* Utrecht and Antwerp: Kosmos, 1991.

Nijhof, Peter and Gerlo Beernink. *Industrieel erfgoed: Nederlandse momumenten van industrie en techniek.* Utrecht and Wormer: Teleac en Inmerc, 1996.

Oddy, Andrew, ed. *The Art of the Conservator.* London: British Museum Press, 1992.

Oei, Loan, ed. *Indigo: leven in een kleur.* Weesp: Fibula Van Dishoeck, 1985.

Ordish, George. *The Living House.* London: Rupert Hart-Davis, 1960.

Pagels, Elaine. *Adam, Eve and the Serpent.* London: Weidenfeld and Nicolson, 1988.

Patterson, J. H. *The Man-Eaters of Tsavo and Other East African Adventures.* 1st ed. 1907. Glasgow and London: Fontana, 1973.

Penninx, Kees. *Beeldvorming over ouder worden.* Houten and Diegem: Bohn Stafleu Van Loghum, 1995.

Pierce, Benjamin A. *The Family Genetic Sourcebook.* New York: John Wiley & Sons, 1990.

Piras, Susanne, ed. *Latrines: antieke toiletten, modern onderzoek.* Meppel: Edu'Actief, 1994.

Pleij. H. *Het gilde van de blauwe schuit: literatuur, volksfeets en burgermoraal in de late middleeuwen.* Amsterdam: Meulenhoff, 1983.

Preston, Douglas J. *Dinosaurs in the Attic: An Excursion into the American Museum of Natural History.* New York: St Martin's Press, 1986.

Purcell, Rosamond Wolff and Stephen Jay Gould: *Illuminations: A Bestiary.* London and New York: W. W. Norton & Company, 1986.

Purcell, Rosamond Wolff and Stephen Jay Gould. *Finders, Keepers: Eight Collectors.* London: Hutchinson Radius, 1992.

Rademaker, L. A. *Crematie en het crematorium te Velzen.* Amsterdam: A. J. G. Strengholt, 1940.

Rammeloo, J., ed. *De huiszwam en andere schadelijke zwammen in gebouwen.* Meise: Nationale Plantentuin van België, 1989.

Rawie, Jean Pierre. *Woelig stof.* Amsterdam: Bert Bakker, 1989.

Reumer, Jelle, W. F. and Kees Moeliker. *Depotographie: bizarre foto's uit het Natuurmuseum-depot*. Photography by Sandi Warnaar. Rotterdam: Natuurmuseum, 1991.

Ricklefs, Robert E. and Caleb E. Finch. *Aging: A Natural History*. New York: Scientific American Library, 1995.

Rietveld, W. J. *Biologische ritmen: een inleiding in de chronobiologie*. Utrecht and Antwerp: Bohn, Scheltema & Holkema, 1984.

Robertson, James. *The Complete Bat*. London: Chatto & Windus, 1990.

Romans, Godiried. *Kopstukken*. Amsterdam: Elsevier, 1961.

Rose, Kenneth Jon. *The Body in Time*. New York: John Wiley & Sons, 1988.

Rose, Michael, R. *Evolutionary Biology of Ageing*. Oxford and New York: Oxford University Press, 1991.

Rosebury, Theodor. *Life on Man*. London: Martin Secker & Warburg, 1969.

Rossi, Paolo. *The Dark Abyss of Time: The History of the Earth and the History of Nations from Hooke to Vico*. Chicago and London: University of Chicago Press, 1984.

Rümke, H. G. *Levenstijdperken van de man*. Amsterdam: De Arbeiderspers, 1938.

Russell, Jefrey Burton. *Lucifer: The Devil in the Middle Ages*. Ithaca and London: Cornell University Press, 1984.

Ryder, M. L. *Hair*. London: Edward Arnold, 1973.

Schäfer, Rudolf. *Der ewige Schlaf. Visages de morts*. Hamburg: Keller, 1989.

Schama, Simon. *Landscape and Memory*. London: HarperCollins, 1995.

Scheepmaker, Nico. *Over Alles*. Amsterdam: Nijgh & van Ditmar, 1991.

Scott, Sir Walter. *Ivanhoe*. 1st ed. 1820. Edited and with an introduction by A. N. Wilson. Harmondsworth: Penguin, 1984.

Segal, Sam. *Flowers and Nature: Netherlandish Flower Painting of Four Centuries*. Amsterdam: Hijnk International, 1990.

Senden, Leo. *Bewoners van krotten en achterbuurten*. Antwerp: Boekhandel de Standaard/ Vlaamsche Boekcentrale, 1936.

Sliggers, B. C. and A. G. van der Steur. *Portretten van Nederlandse "honderdjarigen"*. Haarlem: Teylers Museum, 1989.

Slijper, E. J. *De geheimen van reuzen en dwergen in het dierenrijk*. Leiden: A. W. Sijthoff, 1964.

Smit, Pieter. *Artis: een Amsterdamse tuin*. Amsterdam: Rodopi, 1988.

Smith, Anthony. *The Body*. London: George Allen & Unwin, 1968.

Snijders, J. Th., ed. *Laat seizoen: gedichten voor ouderen*. Houten: Agathon, 1989.

Sparks, John. *The Sexual Connection: Mating the Wild Way*. London: Sphere, 1979.

Stavenuiter, Monique, Karin Bijsterveld and Saskia Jansens, eds. *Lange levens, stille getuigen: oudere vrouwen in het verleden*. Zutphen: Walburg Pers, 1995.

Stearn, William T. *The Natural History Museum of South Kensington: A History of the British Museum (Natural History) 1953–1980*. London: Heinemann, 1981.

Stearns, Stephen C. *The Evolution of Life Histories*. Oxford, New York, Tokyo: Oxford University Press, 1992.

Stewart, Alison G. *Unequal Lovers: A Study of Unequal Couples in Northern Art*. New York: Abaris Books, 1979.

Stoddart, Michael. *The Scented Ape: The Biology and Culture of Human Odour*. Cambridge: Cambridge University Press, 1990.

Strehler, Bernard L. *Time, Cells and Aging*. New York: Academic Press, 1977.

Stubbe, Hannes. *Formen der Trauer*. Berlin: Reimer, 1985.

Stuijvenberg, Willem van. *De toekomst van het ouder worden: praktische gids voor gepensioneerden*. Baarn: Tirion, 1988.

Suzman, Richard M., David P. Willis and Kenneth G. Manton, eds. *The Oldest Old*. New York and Oxford: Oxford University Press, 1992.

Swain, John. *A History of Torture*. 1st ed. 1931. London: Tandem Books, 1965.

Swift, Jonathan. *Gulliver's Travels*. 1st ed. 1726. London: Penguin Classics, 1985.

Thomas, Carmen. *Ein ganz besonderer Saft: Urin*. Keulen: Vgs, 1993.

Thomas, Carmen. *Berührungsängste? Vom*

Umgang mit der Leiche. Keulen: Vgs, 1994.

Thomas, Lewis. *Een brein op een miljoen pootjes.* Amsterdam: Contact, 1987.

Thomson, G. *The Museum Environment.* London: Butterworths, 1978.

Thomas, Keith Stewart. *Living Fossil: The Story of the Coelacanth.* London: Hutchinson Radius, 1991.

Thornton, Peter and Helen Dorey. *A Miscellany of Objects from Sir John Soane's Museum: Consisting of Paintings, Architectural Drawings and Other Curiosities from the Collection of Sir John Soane.* London: Laurence King, 1992.

Thijsse, Jac. P. *Paddenstoelen.* Zaandam: Verkade's Fabrieken, 1929.

Topffer, Rodolphe. *De zonderlinge avonturen van Primus Prikkebeen.* Dutch translation from the French by Gerrit Komrij. Amsterdam: Loeb, 1980.

Toth-Ubbens, Magdi. *Verloren beelden van miserabele bedelaars: leprozen; armen; geuzen.* Lochem and Ghent: De Tijdstroom, 1987.

Trimmer, Eric J. *Rejuvenation: The History of an Idea.* London: Robert Hale, 1967.

Tudge, Colin. *The Food Connection: The BBC Guide to Healthy Eating.* London: BBC, 1985.

Turner, Alice K. *The History of Hell.* New York, San Diego and London: Harcourt Brace & Company, 1993.

Veen, Hanneke van, ed. *Het hergebruikboek: dubbellang plezier van duizend en één dingen.* Boxtel: De Kleine Aarde, 1980.

Vervoorn, Richard, ed. *Noorderkerk Amsterdam: Bouw, interieur, orgel, restauratie, functie.* Amsterdam: Stichting Vrienden van de Noorderkerk, 1992.

Vevers, Gwynne. *London's Zoo: An Anthology to Celebrate 150 Years of the Zoological Society of London, with its Zoos at Regent's Park in London and Whipsnade in Bedfordshire.* London, Sydney and Toronto: The Bodley Head, 1976.

Vogel, Steven. *Vital Circuits: On Pumps, Pipes, and the Workings of Circulatory Systems.* Oxford and New York: Oxford University Press, 1992.

Voute, A. M. and C. Smeenk. *Vleermuizen.* Zwolle: Waanders, 1991.

Waal, Frans de. *Chimpansee-politiek: macht en seks bij mensapen.* Amsterdam: H. J. W. Becht, 1982. Published in English as *Chimpanzee Politics: Power and Sex among the Apes.* Baltimore and London: John Hopkins University Press, 1982.

Waal, M. de. *Dieren in de volksgeneeskunst.* Antwerp: De Vlijt, 1982.

Ward, Peter Douglas. *On Methuselah's Trail: Living Fossils and the Great Extinctions.* New York: W. H. Freeman and Company, 1992.

Warren, Nigel. *Metal Corrosion in Boats: The Prevention of Metal Corrosion in Hulls, Engines, Rigging and Fitting.* 1st ed. 1980. London: Adlard Coles Nautical, 1998.

Weisman, Avery D. *De dood nabij: een psychiatrische studie over de dood en de ontkenning van het sterven.* Bilthoven: Ambo, 1972.

Wetenschappelijke Raad voor het Regeringbeleid. *Duurzame risoco's: een blijvend gegeven.* The Hague: Sdu, 1994.

Whitbourne, Susan Krauss. *The Aging Body: Physiological Changes and Psychological Consequences.* New York: Springer, 1985.

Whitrow, G. J. *Het tijdsbegrip in de moderne wetenschap.* Utrecht and Antwerp: Het Spectrum, 1965.

Wilmink, Willem. *Verzamelde liedjes en gedichten.* Amsterdam: Bert Bakker, 1986.

Wissen, Ben van, ed. *Dodo (Raphus cullatus) (Didus ineptus).* Amsterdam: ISP/Zoölogisch Museum, 1995.

Wood, Gerald L. *The Guinness Book of Pet Records.* London: Guinness, 1984.

Young, J. Z. *An Introduction to the Study of Man.* Oxford: Oxford University Press, 1971.

Zeuner, Frederick E. *Dating the Past: An Introduction to Geochronology.* London: Methuen & Co, 1946.

Zinsser, Hans. *Rats, Lice and History.* 1st ed. 1935. New York: Pocket Books, 1945.

Zomeren, Koos van. *Het scheepsorkest.* Amsterdam: De Arbeiderspers, 1989.

ACKNOWLEDGEMENTS

Page iii: Philippe de Champaigne (1602–1674), *Vanitas*, Musée des Beaux Arts, Le Mans, France; Page viii: Joost Swarte, drawing, 1994; *Page 3*: Anon, *c*. 1640; *Page 6*: Anon, wood engraving, Duitsland, *c*. 1480; *Page 7*: Anon, Duitsland, C15th ; *Page 9*: Rhonald Blommestijn, drawing, 1996; *Page 11*: George Burggraaf; *Page 16*: C. P. Carbo; *Page 18*: Mark Weller; *Page 21*: John Jennens, lithograph, 1862, Mary Evans Picture Library; *Page 24*: G. Riebicke, from Hans Surén, *Mensch und Sonne*. Berlin, 1936; *Page 26*: Charles Louis Clérisseau (*c*.1721–1820), View of the Forum of Nerva in Rome; *Page 28*: Coen Gravendaal; *Page 29*: Jacob Matham after Hendrik Goltzius, 1598; *Page 31*: ABC Press; *Page 32*: Gabinetto dei Disegni delle Stampe, Rome. Inv. nr. 491; *Page 33*: Collection of the Mayor of Regierenden Fürtsen von Leichtenstein, Vaduz; *Page 34*: Kupferstichkabinett, Sammlung der Zeichnungen und Druckgraphik. Inv. nr. KdZ 12918, Staatliche Museum, Berlin; *Page 36*: Sam Wagenaar/NFA; *Page 38*: Arthur Rackham, drawing, 1935, Mary Evans Picture Library; *Page 40*: Jaap Hoogenboom/Natura; *Page 42*: George Burggraaf; *Page 44*: From M. Godwin, *Angels: An Endangered Species*. London, 1990; *Page 45*: Johann Nussbiegel, engaving after G. Dadlebeck, *Vampyrum spectrum*; *Page 47*: Kees Hageman; *Page 52*: Charles Breijer/NFA. A Boer *c*.1938; *Page 55*: Peter Martens/NFA. Leper, Agua de Dios, 1972; *Page 56*: H. Saftleven, drawing 1674–1677; *Page 58*: Albrecht Altdorfer, etching, inv. nr. 1926/1779, Wenen, Albertina; *Page 60*: Gizeh, The Sphinx before 1875; *Page 63*: B Rheims, 1992; *Page 67*: Simon Marsden, Sculpture at Toddington Manor, Gloucestershire; *Page 70*: Robert Harding Picture Library; *Page 74*: Frits van Daalen/ Natura; *Page 76*: M. Wolf/ ABC Press; *Page 78, 79*: Ronald Hoeben, 1997; *Page 80*: From J. Hawkes, *The Atlas of Early Man*. London, 1976; *Page 83*: IFA/Fotostock; *Page 88*: *Left*: Superstock. *Right*: Thomas Howard/Associated Press, 1928, Sing-Sing; *Page 90*: From Gillo Dorfles, *Der Kitsch*. Gütersloh, 1977; *Page 91*: Penny Gentieu/Transworld; *Page 95*: Publicity photograph of the actor Ken Maynard, 1929; *Page 97*: Jacopo della Quercia, *c*.1406–1408, Lucca, San Martino; *Page 98*: Tommaso Giovanni di Masaccio, *Adam and Eve Being Banished from the Garden of Eden*, Brancacci Chapel, Florence; *Page 100*: William Hogarth, *Time Smoking a Picture*, engraving, 1761, private collection; *Page 102*: Rob Huibers/Hollandse Hoogte; *Page 103*: Gerald Davis/Transworld; *Page 104*: Philiberts Bouttats de Jongere, *The Baker of Eeklo*, copper engraving, Atlas van Stolk, Historisch Museum, Rotterdam; *Page 106*: Simone Signoret by Julien Quideau/ABC Press; *Page 107*: Fotostock; *Page 100*: Tim Bobbin, 1773, Mary Evans Picture Library; *Page 111*: Fotostock; *Page 116*: Ad. Mulder, drawing from Bulletin KNOB, 1996–2/3; *Page 117*: Cooper-Hewitt Museum, New York; *Page 118*: ANP; *Page 120*: Jan Steen (1625/26–1679) *De keuze tussen jeugd en rijkdom*, Museum Narodowe, Warsaw; *Page 122*: From W. F. Stearn, *The Natural History Museum at South Kensington*. London, 1981; *Page 125*: From a portrait of Hartsoeker, 1694; *Page 128*: ABC Press; *Page 129*: Tony Stone Images/ Hulton Getty, 1923; *Page 131*: Tony Stone Images/Hulton Getty, 1955; *Page 135*: Michel Gunther/BIOS; *Page 136*: Pier Leone Ghezzi, pen and ink drawing, (1674–75); *Page: 140, 141*: Streekmuseum, Admiraliteitshuis, Dokkum; *Page 144*: From Flammarion, *Hemel n aarde*, engraving by B. C. Goudsmit. Zutphen, n.d; *Page 145*: Peter Maartens/Hollandse Hoogte, Spain 1986; *Page 146*: Jan van Arkel/Natura; *Page 148*: Fred Hazelhoff/Natura; *Page 149*: Publicity photo for the actor Ronald Reagan, Superstock; *Page 154*: From H. Robin, *The Scientific Image*. New York, 1992; *Page 155*: Peter Cushing in *Tales from the Crypt*, Cinerama Releasing; *Page 156*: From A. Destrée,

Geschiedenis van de techniek. Brussels, 1980; *Page 161:* ANP/EPA; *Page 162:* Manchester Museum, Manchester; *Page 163:* Judith Szabo/Hollandse Hoogte, Budapest, 1985; *Page 170:* Anon, Base-relief in wax, C17th. Naples, Congrega di Santa Maria; *Page 172:* Martijn de Jonge; *Page 174, 175:* From *Der dotendantz mit figuren clage und antwort von alle staten der werlt.* Leipzig, 1922; *Page 178:* Christopher Cormack/Impact Photos; *Page 179:* B. Barbey/ABC Press; *Page 182:* From *De begrafenis-moelijkheden in 1945 te Amsterdam.* Amsterdam, n.d; *Page 184:* W. Reiche, drawing, 1874; *Page 185:* From: J. Franke, *Crematie in Nederland 1875–1955.* Utrecht, 1989; *Page 186:* From M. Andrews, *The Life that Lives on Man,* London, 1976; *Page 192:* Siamese Twins, anatomical specimen prepared by Willem Vrolik, C18th, Collectie Anatomisch Embryologisch Laboratorium, AMC, Amsterdam; *Page 193:* Rosamund Wolff Purcell. Anatomical specimen prepared by Frederik Ruysch, C18th; *Page 194:* Glass-plate souvenir, C19th, Collection of D. van Hooff; *Page 196: Left:* Antonio Chichi, Model of the Temple Minerva Modica (c.1750–c.1805) RMO collection, Leiden. *Right:* From Gillo Dorfles, *Die Kitsch.* Gütersloh, 1977; *Page 198:* Mirja de Vries; *Page 201:* From Pierre Boitard, *Études antediluviennes, Paris avant les Hommes.* Paris, 1861; *Page 203:* Lithograph, C18th, Mary Evans Picture Library; *Page 206, 208:* François le Diascorn/NFA; *Page 209:* Rosamund Wolff Purcell; *Page 210:* Collectie Anatomisch Embryologisch Laboratorium, AMC, Amsterdam; *Page 211:* Rosamund Wolff Purcell. Anatomical specimen prepared by Frederik Ruysch, C18th; *Page 213:* W. Kruyt/Natura; *Page 214:* From G. Richter, *Geology for Beginners.* London, 1843; *Page 217:* Martin Harvey/Natura; *Page 218:* From G. Richter, *Kitschlexicon von A bis Z.* Gütersloh, 1985; *Page 219:* Spaarnestad Fotoarchief; *Page 222:* Domenico Feti, *Melancholy,* c.1613, Musée de Louvre, Paris; *Page 227:* Bert Verhoeff; *Page 230:* György Konescni, poster, 1935; *Page 231:* Tony Stone Images/Hulton Getty; *Page 234:* Poster, C19th; *Page 224:* From E. A. Sutton, *The Happy Isles.* London, 1938; *Page 225:* RMO Collection, Leiden; *Page 236:* Science Photo Library; *Page 239:* Reuters; *Page 242:* John Launois/Transworld; *Page 246:* Arthur Rackham, drawing, 1900, Mary Evans Picture Library; *Page 248:* Rudolf Schäfer, from R. Schäfer, *Der ewige Schlaf.* Hamburg, 1989; *Page 252:* George Burggraaf; *Page 253:* Tony Stone Images/Hulton Getty, 1952; *Page 254:* Woodcut C12th/13th; *Page 237:* John Collier, 1881, National Portrait Gallery, London; *Page 260:* Tony Stone Images/Hulton Getty. Four women in Jacobean clothing, England 1929; *Page 261:* Frans de Waal, from F. de Waal, *Chimpanzee Politics,* John Hopkins University Press, 1982; *Page 264:* Spaarnestad Fotoarchief; *Page 266:* Henricus Hondius, engraving, 1649.

Acknowledgement for the epigraph is made to Ann Rae Jonas and The Word Works for the quotation from "Structures" which is published in *A Diamond is Hard but not Tough,* by Ann Rae Jonas (Washington, DC: The Word Works, 1998).

INDEX